MAKING THE MOST OF YOUR TRACTOR

Contents

		Page
Foreword		i
Introduction		ii
SECTION 1	**How to work safely**	1
	1 General safety (PY1 A1)	2
	2 Emergency aid in case of chemical poisoning (COM 7.1)	15
	3 Lifting and carrying (COM 1.1)	18
SECTION 2	**How to operate your tractor**	27
	4 Daily checks (M1 A1)	28
	5 Familiarisation with controls (M1 A2)	34
	6 Driving straight forward and in reverse (M1 A3)	40
	7 Driving around the farm (M1 A4)	43
	8 Attaching mounted implements to the three-point linkage (M1 B1)	46
	9 Hitching trailed machines (M1 B2)	51
	10 Attaching pto shafts (M1 B3)	56
	11 Connecting machines to the external hydraulic services (M1 B4)	59
	12 Manoeuvring with mounted machines (M1 B5)	62
	13 Manoeuvring with a two-wheeled trailer or trailed machine (M1 B6)	64
	14 Jacking up tractors and wheeled machines (M1 C1)	67
	15 Fitting wheel and frame weights (M1 C2)	70
	16 Altering wheel track width (M1 C4)	73
	17 Water ballasting tyres (M1 C5)	75

Contents

		Page
SECTION 3	**How to maintain your tractor**	81
18	Servicing tractors (excluding engines) (M1 D1)	82
19	Engine lubrication (M4 A1)	86
20	Servicing cooling systems (M4 A2)	89
21	Servicing vehicle electrical systems (M4 B1)	92
SECTION 4	**How to carry out field operations**	95
22	Operating a fore-end loader (M5 H1)	96
23	Ploughing with a mounted right-hand mouldboard plough (M6 A1)	102
24	Ploughing with a mounted reversible mouldboard plough (M6 B1)	113
25	Pre-season preparation of seed drills (M7 A1)	131
26	Post season maintenance of seed drills (M7 A5)	143
27	Preparing field crop sprayers (M9 D1TG)	150
28	Calibrating field crops sprayers (M9 D1TA)	157
29	Crop spraying (M9 D2)	161
30	Maintaining reciprocating knife mowers (M10 A1)	169
31	Operating reciprocating knife mowers (M10 A2)	176

Foreword

British agriculture is amongst the most efficient in the world. It employs highly-skilled operators using some of the most advanced and up-to-date agricultural machinery available.

One of the most important reasons for this achievement is the high priority given to training by the agricultural industry in Britain. Successful methods of training for the many and varied farm operations have been developed over the years by colleges and institutes supporting agriculture. A special feature of the training effort for farmers and their staff in Britain is the Agricultural Training Board. The Board, through its experienced staff, has produced a wide range of trainee guides in support of its courses for those working or intending to work on farms. These guides are highly rated by British farmers, and several have been used as the basis for this book. The guides may also be found in use in many countries around the world, especially where effective mechanisation undertaken by well trained operators provides one of the most direct ways of solving the pressing problems of food production. It therefore follows that this book will be of interest and practical value to those connected with agriculture in Britain and overseas.

I commend this book to all trainees and instructors concerned with tractors and associated machinery. It will help them to get the best out of costly equipment which requires a high level of operator skill for efficient and safe operations.

Brian May
Silsoe February 1985

MAKING THE MOST OF YOUR TRACTOR

Introduction

The aim of this book is to improve the standard of operation of tractors and associated machinery in many parts of the world. The operation of tractors, like that of all machinery, involves more than climbing into the seat and driving off; it also means maintaining the machinery to stop it wearing out and breaking down too often; making efforts to avoid injuring people and animals; and preventing damage to equipment and crops. The book aims to give practical guidance on all these essential parts of the tractor driver's work.

The book has its origin in the series of trainee guides produced by the Agricultural Training Board (ATB) in the United Kingdom, guides which were designed for use by people attending the ATB training courses. The ATB was set up to train people actually working on farms, using instructors with plenty of agricultural experience, backed up by written reference material organised on a topic basis. The ATB guides thus contain a great deal of practical and supportive information and help for the tractor-operator — and this help should be useful to the operators of tractors and associated machinery, no matter in what country they are working. Most of the guides are written in general terms and are planned for use with ATB's experienced instructors — and with the very specific manuals provided by manufacturers for individual machines. These 'operator's manuals' are always provided with new machines but unfortunately they seldom end up in the operator's hands. It is very much hoped that everyone who reads this book will then insist on having access to the manuals for the equipment they will be using, or for which they are responsible. If you do not have the relevant operator's manual there are three steps which you can take:

1. ask your manager to get hold of a copy for you;
2. ask the dealer who supplied the machine to provide one;
3. write to the manufacturer of the machine asking for one.

Those using machines should always remember that they are responsible both for the machines and for the safety of other people — and safety is stressed throughout this book. This publication is generally relevant to any farming system, but various British regulations and official publications are mentioned in it — for the benefit of those readers who can get hold of them. Therefore very few alterations have been made to the original text. Enquiries about the original guides — and some of those referred to in the text are not included in the present volume — should be sent to:

Kingsley Bungard,
Agricultural Training Board,
32–34 Beckenham Road, Beckenham,
Kent BR3 4PB, United Kingdom

The text presented here will, it is hoped, be of equal use to teachers and to trainees. Volumes are planned on cattle, sheep, poultry and land and building management; and readers suggestions are welcome on titles of other volumes or individual topics that could be covered such as more detailed tractor maintenance, disc ploughs, cultivating implements, threshers and pumping equipment. The editor and the publisher would be very glad to have your suggestions about what the book should include and what could be left out of this volume. Please write to us at the above address to tell us your ideas.

How to use the book:

First you should read the introduction to the section which interests you, and then find the relevant topic guide in that section. Read the paragraph headed 'Your Aim' at the beginning of each topic. This will tell you clearly what you can hope to achieve if you follow the instructions given. Next, look at the section 'You will need' to find out the necessary equipment for the job. Do not be discouraged if you do not have all the equipment to start with; make use of what you have but try to persuade your manager that the items listed in the guide are necessary to use the machine to full advantage. You should also remember, of course, that the manual provided by the manufacturer is the most important guide to consult. Make sure you read the items marked △ — these are important safety points and it is vital to understand and follow them to prevent accidents. Every day in some part of the world accidents occur simply because people don't take notice of or remember important safety points.

You will notice that the table of contents lists the code numbers of the original guides so that you can trace the references in the text.

Finish reading the rest of the topic and try to apply it to your own situation — with the help of the manufacturer's operator's manual. When you read something you don't understand don't ignore it, try to find out what it means. You could ask more experienced tractor drivers in your area, or your local extension officer. Whatever happens, do not give up — keep asking questions.

When you have understood the instructions and carried them out, read the topic again — it should already be clearer to you having put it into practice. Now try to explain to someone else all the points under the heading ' You should understand' which is at the end of each topic. If you can do this well and answer the other person's questions about the subject, you are on the way to becoming a better tractor operator!

Acknowledgements

The undersigned has brought together the work of the original authors of the ATB Trainee Guides; and would like to give particular thanks to the Director and members of the Graphics, Photographic and Overseas Divisions of the National Institute of Agricultural Engineering, and to Kingsley Bungard of the Agricultural Training Board for much advice and encouragement, as well as for the use of the original material itself.

September 1984

R J Sims

SECTION 1: How to work safely

This section comes first to make sure that you read it! It is hoped that you will read it more than once, and that you will refer to it when you observe unsafe activities taking place around the farm — it is meant to make you very conscious of safety in all situations.

Compared with other industries, agriculture has a bad record for death amongst the work-force. In Britain alone, with only about 1.3 million people employed in agriculture:

- a child dies every two or three weeks
- an adult dies every three to four working days
- an adult is seriously injured every working hour
- machinery is the biggest single killer

Before you handle a crop sprayer read 'Chemical Poisioning' section again so that you know what action to take if some is affected by the chemical. Of course, positive awareness on your part of the possibility of chemical poisoning makes it far less likely that anyone will be poisoned!

The 'Lifting and Carrying' guide is included because most agricultural workers suffer from bad backs at some time in their lives. Awareness of the causes of back-strain will help you to avoid them.

1 GENERAL SAFETY

Introduction

1. This guide lists the safety precautions relevant to the day-to-day running of a poultry site. It is not aimed at one specific level of employee, as some points obviously concern only management while others are of importance to all workers in the industry. It has been left to the individual reader to identify areas in which he or she has responsibilities for safety and to ensure that the recommendations made are applied to management of the site and normal work routines.

2. Although every attempt has been made to cover the main points, the list is by no means exhaustive and users of the guide are advised to make additions applicable to situations within their control.

3. Where training in safety is to be provided, it is recommended that safety points are not taught in isolation but are included, with emphasis, in instruction in the various activities to which the points relate and/or are incorporated at an appropriate stage into a training programme.

Note:
It must be emphasised that the purpose of this guide is to list safety precautions. It is not intended as a statement or interpretation of any law.

1. Maintaining a safety policy

Ensure that each member of the site staff is made aware of:

- the safety policy of the business
- the requirements of the Health and Safety at Work Act and the various regulations relating to fire prevention, using machinery, using chemicals, lifting heavy weights and the general safety of work places
- what to do in the event of a fire
- what to do if someone is injured, receives an electric shock or comes into contact with a dangerous chemical substance
- the name of the individual(s) to whom to report in the event of an emergency or hazardous situation
- the hazards to safety and health which exist.

Keep a suitable number of first aid kits on the site and make sure that they are stocked with usable materials at all times and not misused.

Ensure that members of staff have access to and make full use of:

- personal washing and toilet facilities, including hot water and non-abrasive skin cleaning materials (especially after using chemicals, cleaning-out or handling dead birds or rodents)
- an appropriate range of types and sizes of clean, comfortable, protective clothing (for general use and especially in situations involving dust, fumigants, disinfectants and insecticides) including waterproof boots and dust masks which cover the nose and mouth.

Arrange for protective clothing to be cleaned regularly. Dirty clothes can cause skin irritations and can even be a fire risk.

Ask the local fire service for advice on:

- the siting, maintenance and operation of fire-fighting appliances
- escape routes and access
- storage of highly-inflammable materials (e.g. fuel oils, liquefied pressure gases, paints and preservatives)
- relevant fire prevention regulations
- regular inspection by, and liaison with, the fire service.

Make sure that members of the site staff adopt a policy of reporting any accident, safety hazard or health hazard, however minor it may seem, to the appropriate individual as soon as possible after its occurrence or discovery.

Arrange for all members of the site staff to be immunised regularly against tetanus.

Discourage:

— 'short cuts'

— skylarking

— complacency about hazards

— lack of consideration for other people and property.

Keep an accident record.

If an operator is working alone on a site, make sure that:

— somebody else knows that the operator is working alone and knows his or her expected time of arrival home, at an office or at another site

— a check is made that the operator concerned has arrived safely as planned, or has reported in

— failure of the operator to arrive or report as planned is investigated *immediately.*

2. Moving around the site on foot

(a) Lifting and carrying

Always use a safe method when lifting and carrying heavy objects. Get help if necessary.

Do not try to stop a heavy object which begins to fall while it is being handled. Let it fall, or serious injury may result.

(b) Using ladders

When using a portable ladder or stepladder:

— make sure that the ladder is in good repair and is strong enough and long enough for the job (i.e. the top of a long ladder must reach beyond where you have to get on and off it)

— stand the ladder or stepladder on a firm, level surface and make sure that a long ladder is at a safe angle (i.e. not less than $30°$ from the vertical) or is tied securely to a part of the building

— whenever possible, get somebody to steady the bottom of a long ladder for as long as you are on it

— move slowly and carefully when ascending and descending any ladder; hold on to the rungs and not the stiles

- do not try to reach further than arm's length or lean out to one side while working on any ladder.

Take extra care when climbing fixed, vertical ladders on bulk bins. Wear non-slip footwear and carry out the inspection of the inside of a bin only when another person can watch you from the ground.

Maintain all ladders and stepladders in a safe condition.

When carrying a portable metal ladder, take care not to contact overhead power lines.

(c) Working on a roof

If it is necessary to work on a poultry house roof, use crawling boards or soundly constructed roof ladders and move carefully.

Take care, when on a roof, not to contact overhead power lines.

(d) Using droppings pits or disposal pits

Gases generated by droppings/slurry pits and covered pits used for the disposal of dead birds can be fatal to man. Also, there is the possibility that one of the gases released by a droppings/slurry pit is flammable. Therefore:

- do not smoke or use a naked light near a droppings/slurry pit
- if it is necessary to enter any pit, thoroughly ventilate the pit first
- enter a pit only if wearing an approved type of respirator, a safety harness and a life-line held by two other people who can see you at all times and who know that they must pull you out at the slightest sign that you are in difficulty
- when a droppings/slurry pit is being emptied, make sure that the area is well ventilated and do not stand on or near the slats
- do not allow droppings/slurry to rise to a level which is less than 30 cm below the slats or covers
- keep all pit covers in a safe condition and make sure that no cover can fall into its pit
- if a pit cover is lifted off for any reason, make sure that it is replaced as soon as possible
- if a pit has to be left uncovered, see that the opening is securely fenced or guarded and that the area is well lit during the hours of darkness.

(e) Keeping floors safe

Remove slippery areas on floors by thorough washing and brushing.
Put down sand temporarily on greasy areas which will not wash off easily.

Remember that rubber boots slip easily on smooth, wet concrete.

Make sure that wooden floors are kept in a good state of repair.

(f) Working near machinery

When working near moving machinery:

— remove or fasten back loose or flapping articles of clothing

— fasten back long hair

— remove wrist-watch and all items of jewellery, as both can become caught on projections or cause an electrical short circuit, leading to severe burns.

3. Moving around the site in vehicles

(a) Driving around the site

Observe the safety precautions relating to driving tractors, especially:

— remove mud, grease and any other slippery matter from footwear and footplate before mounting a tractor

— operate all controls only from the driving seat

— do not carry passengers on a tractor, a draw-bar or a loaded trailer

— make sure that any passengers carried on an empty trailer are seated in the middle of the trailer floor

— before dismounting from a tractor, put the gear lever in neutral, apply the parking brake, lower any raised attachment to the ground and switch off the engine.

Observe any speed limits.

Drive with extra care when:

— other vehicles (e.g. catching team's lorries; cleaning-out machinery) are working in the area

— children are likely to be present.

As a general rule, do not allow children on to a site while work is in progress.

Do not allow any unauthorised person to drive any vehicle or operate any machinery.

(b) Using vehicles in buildings

If using a hydraulically-operated loader or trailer in a poultry house, take extra care where any part of the building is lower than the machine's upper limit of lift.

Tie back hanging lights if these are likely to be damaged.

When using a tractor (or any internal combustion engine) inside a house, open all doors, air inlets and fan shafts and run fans at high speed to remove toxic exhaust fumes.

(c) Using a fore-end loader

Do not use a fore-end loader as a crane to left heavy objects or to lift a machine so that somebody can work underneath it.

If a tractor stalls with a raised fore-end loader in contact with overhead power-lines, do not climb down from the machine but jump well clear. Keep well away from the tractor and make sure that nobody else goes near it until the power in the lines has been switched off.

(d) Maintaining vehicles

Make sure that all vehicles used on the site are maintained in a safe condition, particularly with regard to guards and jacks.

Before working or reaching under a vehicle which is raised on jacks or on hydraulics, make sure that it is safely propped up with additional blocks or stands and, where appropriate, that wheels are chocked.

4. Ensuring the safety of equipment, buildings and roads

(a) Using tools and portable items of equipment

As a general rule, be tidy. Untidiness can cause injuries and might mean that an escape route is obstructed in an emergency.

Do not allow cables, hoses or ropes to lie where they can trip people. Whenever possible, arrange them so that they are out of the way while being used and stow them away neatly immediately after use.

Do not use any tool or item of equipment for any purpose other than that for which it is designed.

Return each tool and item of equipment to its recognised storage place immediately after use.

(b) Maintaining fire-fighting equipment

Make sure that an adequate number of fire-fighting appliances of the appropriate types are:

— positioned around the site, as recommended by the fire service

— accessible at all times

— checked and serviced as often as is recommended.

Ensure that all members of the site staff know how to use the various fire-fighting applicances and for which type of fire each type of appliance is suitable.

(c) Using lifting gear and suspension systems

Arrange for all suspension systems (e.g. as used for suspended automatic feeders or canopy brooders), support structures (e.g. tank supports or water or bulk bin towers) and monorail systems to be checked for safety at least once a year.

Find out what the safe working loads of lifting gear and various structural supports are and make sure that they are never exceeded.

Avoid overloading food skips in an attempt to save time.

(d) Maintaining machinery

Make sure that all moving parts of machines are adequately guarded, especially side-wall fans.

Arrange for all machinery and complex equipment to be serviced by a qualified fitter as often as is recommended by the manufacturer or supplier.

Unless you are qualified to do so, never attempt to test or repair any electrical system or equipment. Send for a skilled electrician (see also Section 5).

Before work is carried out on a machine, see that the area is well lit and, if appropriate, adequately ventilated.

(e) Using gas brooders

Before lighting a gas brooder in a poultry house littered with fresh shavings or straw, clear the litter away from directly under the burner.

Clean accumulated dust from gas brooders as often as is recommended by the brooder manufacturer. Take care when cleaning brooders in deep litter houses that pieces of burning carbon do not drop from the burners and start a fire.

Check regularly that a house heated by gas brooders is adequately ventilated and that the brooder burners are working correctly. Incorrect burner operation and/or inadequate ventilation can lead to carbon-monoxide (CO) poisoning.

If you begin to feel drowsy in a house which is heated by gas brooders, leave the building immediately and as quickly as possible by the nearest exit. Arrange for the burners to be checked for CO generation.

Note: Carbon monoxide generation by gas brooders represents a serious hazard because it is lethal but generally unexpected. All members of the site staff must be made aware of the danger. (Metering devices are available and can be installed to detect the presence of carbon monoxide in a house.)

(f) Storing materials in buildings

Do not allow smoking near stored inflammable materials (e.g. straw, shavings).

Take care not to overload walls and upper floors with stored materials (e.g. bagged feedingstuffs, wood shavings).

Remember that some materials, when stored in bulk, are a serious fire and/or vermin hazard. Ask your local Fire Service/Pest Control Officer for advice where necessary.

(g) Protecting fuel tanks

Do not allow smoking near fuel stores.

Keep the ground free of vegetation for a distance of at least 3 m around each storage tank containing liquefied pressure gas or fuel oil.

Avoid weedkillers based on sodium chlorate, as they are a fire risk. Use instead, a preparation based on atrazine, simazine or 2,4,5,–T.

Make sure that all members of the site staff know where the gas/oil tank isolating valves are and how to operate them.

Report a leak from a fuel tank as soon as it is discovered. Where possible, put down sawdust, sand or soil to soak up the leaking material.

(h) Maintaining roads and access routes

Arrange for the surfaces of roads and areas used by vehicles to be maintained in a safe condition.

Keep vehicle access routes clear at all times.

Regularly trim hedgerows and verges near the site exit so that vehicles moving out on to the public road have an unobstructed view *in both directions.*

Make sure that overhead power lines are securely fixed and will not be contacted by vehicles or by anybody climbing a bulk bin ladder.

5. Using and maintaining electrically-operated equipment

Clean accumulated dust from electrical equipment at least once a week.

Connect portable items of electrical equipment to the main power supply only by:

— a cable which is of the correct type and in good condition (e.g. not frayed or cut)

— a sound plug of the correct type and containing a fuse of the correct rating.

Before connecting portable equipment to the mains, make sure that it will be used at the voltage for which it is designed. Connect it through a transformer if necessary.

Replace a damaged or faulty plug or cable as soon as it is discovered.

Unless you have been trained to do so, never attempt to test or repair any electrical system or equipment. Send for a skilled electrician.

Always switch off the mains power supply and send for a skilled electrician if:

— failure of electrical equipment is due to more than a faulty plug or a blown fuse

— a fuse blows immediately after it has been mended.

Before carrying out any mechanical repair or adjustment on any equipment which is powered by electricity (e.g. auger; automatic feeder), switch off the mains supply to the equipment and withdraw the fuse. If the work is to be done out of sight of the switch/fuse box, hang a warning notice on the switch (e.g. 'Work in progress on feeder chain. Do not switch on feeder').

Always disconnect a portable electrical appliance from the mains supply when it is not in use, by removing the plug from the socket. Check that this has been done at the end of the day.

Avoid touching a switch with a wet hand or touching a wet switch.

Do not rely on gumboots to provide an effective insulation against electric shock.

Clearly label each main switch to show:

— the equipment which it controls

— the 'on' and 'off' positions.

Before washing down a poultry house, especially if a pressure washer is to be used:

— switch off the power supply to the house being washed, if possible

— connect an electrically-powered pressure washer to a point in a neighbouring house, if possible

— fix a waterproof covering to any electrical equipment or fitting which is near those parts of the house where water will be applied.

Never allow water, from whatever source, to contact electrical equipment or fittings.

Before removing a dud light bulb or tube for replacement, isolate the row of lights containing the dud, if possible, by switching off the power supply and withdrawing the fuse.

When fitting a new bulb or tube, make sure that:

— there is no power to the socket

— the bulb/tube is dry (a wet lamp is likely to explode as it heats up).

If it is necessary to fix up a temporary light, make sure that, as well as being connected to the power supply by the correct type of cable and plug, it is not in a position where it is likely to be an electrical or fire hazard.

6. Cleaning out poultry houses

If using a pressure washer, never direct the jet at other people working nearby. (See also points about electrical equipment and fittings in Section 5.)

Always check and follow the manufacturer's instructions for mixing and applying fumigants, disinfectants and insecticides. (Some substances, although not covered by any statutory provisions, can be hazardous to operators. Manufacturers may recommend that protective clothing is worn while mixing and using their products.)

After fumigating a house:

— make sure that everybody who has been working in the house is seen to leave the building and can be accounted for

— lock all doors to the house and remove the keys to a safe place

— fix a warning notice to each door, giving details of the chemical used and any safety precautions recommended by the manufacturer (e.g. forbidding entry, unless protective clothing is worn, until 12 hours after fumigation or until after 6 hours, followed by 1 hour's thorough ventilation)

— make sure that the warning notice is taken down as soon as it is safe to re-enter the house.

If using a fumigating system involving heaters, place each heater on a clear area of concrete floor at least 3 m away from any inflammable material (e.g. wood shavings).

Thoroughly wash your hands, face and any other areas of exposed skin immediately after:

— any cleaning-out operation (e.g. removing poultry litter)

— using any chemical substance.

Avoid the use of abrasive materials (e.g. pumice; scouring powder) when washing the skin. Use only an approved skin cleanser, if necessary, followed by soap and hot water.

7. Handling livestock

Do not underestimate the ability of poultry breeding stock to inflict injury. Handle adult birds carefully, especially geese, turkeys and the male breeders of any species.

Dispose of dead birds and dead rodents as soon as possible after discovery/collection.

Avoid handling dead rodents with your bare hands. Use rubber gloves or a fork/shovel.

Use a disposal system which will ensure that no carcase, whether bird or rodent, can be retrieved or unearthed by dogs or vermin.

Wash your hands and arms thoroughly immediately after handling any dead bird or animal, however safe the method of handling may seem.

When preparing a poultry house for depopulation make sure that any hazards to catchers are:

— removed

— shielded

— sufficiently lit to prevent injury according to what is possible.

Make sure that a lorry-load of full portable poultry crates is securely fastened to the vehicle and not likely to shift or collapse in transit. Check that ropes are in good condition.

8. Storing and disposing of materials

(See also notes in Section 1 about storing inflammable materials and in (f) and (g) of Section 4 about storing materials in buildings and protecting fuel tanks.)

(a) Storing chemical substances and veterinary materials

Store any chemical substance only in the supplier's original container, which should be clearly labelled and tightly closed.

Keep all dangerous substances (e.g. pesticides and fumigants) and veterinary materials under lock and key when not in use and keep the key in a safe place known and accessible only to authorised members of the site staff.

Record the use of chemical substances (e.g. name of material, quantity used, purpose, date).

(b) Disposing of chemical substances

Dispose of any substance which cannot be identified with any certainty. Ask your local county waste disposal authority for advice on safe disposal.

If plastic containers have been burned, bury the remains in a remote part of the site and at least 0.5 m under the ground.

(c) Disposing of dangerous objects

Dispose of broken glass, old light bulbs, old or broken fluorescent tubes, discarded injection needles and old scalpel blades in a manner which will prevent subsequent injury. If possible, place such items in some form of container or wrapping before disposal, making sure that they will not be accessible to children or livestock.

(d) Burning inflammable waste (e.g. cardboard chick boxes)

Choose a place at least 50 m from buildings, fuel supplies, crops, trees, bushes, fences and hay/straw stacks. Avoid areas where sodium chlorate weed killer has been used.

Burn only when there is no wind (e.g. preferably in the early morning).

Burn waste material in small batches at a time.

Always stay with a fire and ensure that it does not spread through grass, stubble, etc.

When all the material has been burned, douse the embers with water and make sure that the fire is completely out before leaving the area.

2 EMERGENCY AID IN CASE OF CHEMICAL POISONING

1. Introduction

This guide, which has been approved by the Health and Safety Executive, gives information on what should be done in the event of chemical poisoning.

Even if you have not been trained in *First Aid*, there is a lot you can do to reduce suffering and even save the patient's life, if you follow the procedures shown in this guide.

2. Symptoms

Below is a list of symptoms from which you, or another person, might be suffering as a result of chemical poisoning:

— headache

— tiredness

— excessive sweating

— thirst

— sickness

— stomach pains

— convulsions

— shortness of breath

— rapid heart beat

— confused thoughts and vision

— lack of muscle co-ordination.

3. Action to be taken

If you yourself feel any of the symptoms mentioned in section 2, get help immediately.

If you notice someone else displaying any of those symptoms, take the following action*:

— stop person working

— take person into shelter, away from chemical

— remove all protective and other clothing contaminated with chemical

— try to get patient to rest

— if patient is semi-conscious or unconscious place him into recovery position shown below

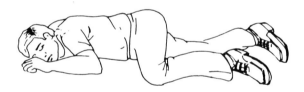

— try to identify the chemical which has been used (see Section 8)

— GET HELP.

* If someone in a glasshouse or other enclosed building appears to be suffering from chemical poisoning, BEFORE YOU ATTEMPT TO EFFECT A RESCUE, put on full protective clothing, including a respirator. (Ideally, this equipment should be close at hand.)

4. Convulsions

If the patient is suffering from a convulsion:

— DO NOT LEAVE PATIENT

— *do not* try to restrain patient's movements, other than preventing him from falling or banging his head

— if possible, remove patient's false teeth.

When the convulsion has subsided, take the patient away from the spray area:

— remove all contaminated clothing

— thoroughly wash all contaminated skin, using soap and water

— keep patient warm

— attempt to identify chemical which has been used

— GET HELP.

5. Breathing ceases or weakens

If patient's breathing ceases or weakens:

— ensure breathing passages are clear

— remove any false teeth

— start artificial respiration immediately, (mouth-to-mouth method should *NOT* be used if patient has *swallowed* any chemical – see Section 6)

N.B. You would be better able to do this if you have been trained.

6. Swallowed chemicals

If the patient has swallowed any chemical, *do not* make him sick. If vomiting does occur, collect a sample and place it into a glass container, which should be kept for examination. If a chemical is corrosive (such as hydrofluoric acid) give the patient milk to drink to dilute the corrosive effect. Get the patient to hospital as quickly as possible.

7. Eye contaminated with chemical

If you suspect that the patient's eye has been contaminated with chemical:

— make patient blink his eye under running drinking water, or flush eye with drinking water for at least 15 minutes, as shown below

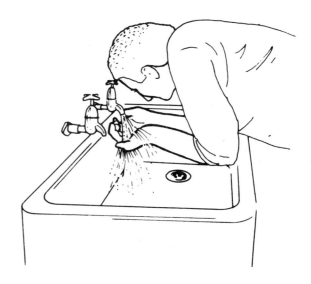

— make sure that uncontaminated eye does not also become contaminated

— *do not* let patient rub contaminated eye

— cover patient's eye with an eye pad from First Aid kit and arrange for patient to receive medical treatment.

8. Dinitro chemical

If patient has been using spray based on a Dinitro chemical, for example:

— BASF DNBP Amine

— Dinoseb Amine Weedkiller M&B

— Farmon DNBP (Amine)

— Supersevtox

— Chafer Dinosol

take the patient away from the source of contamination:

— remove all contaminated clothing

— keep patient:

  — lying flat

  — at complete rest

  — in shade

  — in a cool current of air; if necessary, fan patient

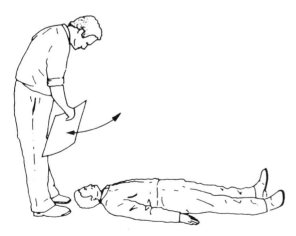

— sponge patient's face with cold drinking water

— try to re-assure patient.

9. If you are also contaminated

If as a result of helping the patient, you yourself become contaminated, remove all contaminated clothing, wash thoroughly and seek medical advice.

10. Transporting the patient to hospital

It is always advisable to call an ambulance, but if you take the patient yourself, put him in the recovery position in the vehicle to avoid inhalation of vomit and, if possible, arrange for another person to accompany you to look after the patient.

REMEMBER

The action you take could save someone's life.

The chances of your saving someone's life is greater if you have been trained in First Aid.

SO IF YOU HAVE NOT, THEN GET TRAINED!

3 LIFTING AND CARRYING

Your aim

is to be able to:

— plan the lifting of any article within your individual capacity

— lift articles within your capacity, using the correct method

— plan and carry out a two-man lift, using the correct method.

1. General points of lifting and carrying

All safe lifting depends on:

— planning the lift

— knowing and testing the weight before attempting to lift it

— knowing your own capabilities

— starting in the correct position

— using your thigh muscles and not your back muscles

— using a correct grip.

Do not attempt to lift or move heavy objects if you have previously suffered from back injuries or back troubles of any kind.

Always test the weight of a heavy object (if the weight is not known) before attempting to lift or move it.

Get help if you think that you will not be able to lift or move an object unaided or if you know that a sack or bag and its contents weigh more than 81kg/180lbs. If help is not available, do not attempt the job alone.

Before handling heavy objects, stop and plan your movements carefully.

Whenever lowering your arms, bend your knees.

Avoid twisting your body when lifting or carrying heavy objects.

2. The correct hold

Grasp objects with right-angled edges (e.g. boxes) using:

— palm of hand

— whole length of fingers and thumb, not finger tips.

Grasp other objects using as much of your hand as possible.

NOTE:

Using only finger tips or any insecure grasp can lead to:

— object dropped, causing injury

— damage to tendons and muscles.

3. The correct lifting position

Body natural and relaxed.

Feet placed to give good balance:

— about 500mm apart, depending upon height

— one foot in front of the other, pointing in direction of movement.

Legs bent.

Chin tucked in.

Back naturally straight.

Arms held to side of body.

Correct grasp (see Section 2).

Thigh muscles used instead of back muscles.

Object lifted with smooth, rhythmical action, not jerked.

Avoid twisting your body while lifting an object and when carrying.

4. The danger of the unsafe lifting position

Body is unbalanced.

Risk of:
- injury to spine
- hernia.

Body weight added to weight of object.

The following are examples of using the correct method.

5. Lifting boxes

Adopt correct position and grasp (see Sections 2 and 3)

Use your body weight, not muscles, to tilt box.

Avoid twisting your body whilst lifting box and when carrying.

6. Lifting paper/plastic sacks from flat on the ground

Use correct position and grasp (see Sections 2 and 3) to lift bag to standing position. (Grip on bag can be increased by moistening hands.)

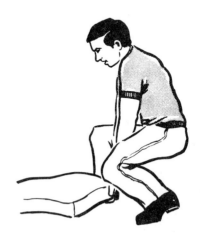

Take one step forward with your rear leg as bag is lifted to standing position.

Use your body weight not muscles, to tilt bag on to one corner.

Use correct position and grasp for second lift — one hand under raised corner.

7. Lifting a draw-bar

NOTE: A draw-bar jack must always be used where one is fitted.

Use correct position and grasp.

Body faces the direction the draw-bar is to be moved.

8. Lifting a large object without handles — two men

Use correct position and grasp.

One person calls time.

Both men lift at the same time and speed.

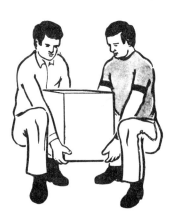

9. Lifting a large object with handles (e.g. a churn) — two men

Outside hands used to lift.

Each man's inside hand placed on other man's shoulder.

Inside hands push as outside hands lift.

Time called by one man (e.g. 'one, two, three, Lift'), to ensure smooth, even lift.

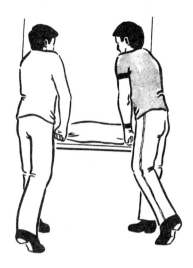

10. Lifting and loading sacks of more than 81kg/180lb weight — two men

A. Direct lift to waist height.

Corners of bag formed into 'ears' for gripping.

Use correct position and good grasp by each man.

Outside feet pointed towards trailer/platform.

B. Swing lift to above waist height.

Corners of bag formed into 'ears' for gripping.

Use correct position and good grasp by each man.

Each man faces away from trailer/platform, with one foot pointed away from direction of lift.

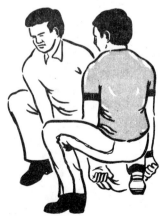

Time called by one man, to ensure smooth, even lift.

Bag lifted and swung away from trailer/platform.

Each man turns to face trailer/platform and follows through with outside leg as bag is swung back.

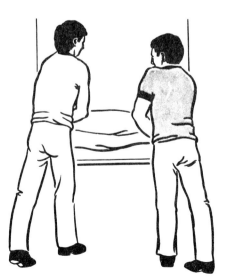

C. Lifting with stick — two men.

Grasp stick with outside hands.

Adopt correct lifting position.

Outside feet pointing in direction of lift.

Place stick under end of sack nearest trailer/platform.

Grasp other end of sack.

Form corners of bag into 'ears'.

One man calls time.

Lift sack, placing end with stick on trailer.

Lift and push other end of sack up and over onto trailer.

11. Handling a drum

A. Raising to upright.

Adopt correct position with your left foot forward beside drum.

Raise drum till weight starts to change.

Move your right hand to front edge of drum. Step forward with your right foot.

Lower drum to upright position, using your left leg as counter-weight.

B. Lowering drum from upright.

Reverse the procedure in Section 11A, using your near leg to assist in tilting drum before changing your hand position to assist in lowering.

Remember

- to use the correct position and grasp for every lift

- to stop and plan your movements before lifting any object

- to get help when you have to lift or move sacks or bags which weigh more than 81kg/180lb, or which are too heavy for you to handle comfortably

- that incorrect lifting can lead to serious and permanent injury.

You should understand

- why your neck and back can be seriously and permanently injured by incorrect lifting

- why, if you bend to lift anything, approximately 35kg is added to the weight of whatever you are lifting

- why planning your movements before lifting and carrying heavy objects is important

- why you are forbidden by law to lift a sack or bag weighing more than 81kg/180lb without help

- why it is illegal for an agricultural worker under the age of 18 to lift, carry or move a load heavy enough to be likely to cause him an injury

- why it is foolish and dangerous to 'show off' by lifting weights which are heavier than you can comfortably handle.

If you do not understand all these points, ask your instructor, manager, employer or course instructor for advice.

Further experience

You will only become proficient by practice. Try to get experience as soon as you can by taking every opportunity offered by your employer or manager to follow up the training you have received.

Remember that the Board and your county agricultural college offer a wide range of courses in practical skills, together with associated knowledge which will help you understand more about the work you are doing.

SECTION 2: How to operate your tractor

This is the major section of the book and takes the reader through the steps of learning to drive a tractor.

If he can already drive a tractor he should use the guides to check on his own standards of operation. He will almost certainly discover that he has developed some bad habits. Six of the tractor driver's most common faults world-wide are:

— driving with his foot resting on the clutch pedal
— using his foot throttle in field work
— cluttering his foot plates with links, pins, or tools
— leaving his brake pedals unlatched
— failing to adjust his tyre pressures to suit the work
— leaving his fuel tank empty overnight

and many others which need a conscious effort to correct. A good standard of operation will not only result in better work in the field, but also in less wear and tear on the tractor and on the driver.

The section covers all the basic elements of tractor driving with a variety of attachments. It is recommended that the reader works his way through each guide carefully, practising each operation described until he can do it safely and accurately. This is the way to become a good tractor operator.

4 DAILY CHECKS

Your aim

is to be able to:
- carry out all necessary daily checks on
  the engine lubrication system
  the cooling system
  the wheels and tyres
  the lighting
- lubricate the tractor as necessary
- check the fuel system and fill the tank
- carry out all tasks in accordance with the manufacturer's recommendations
- take precautions to minimise risks of injury to all persons and livestock
- take precautions to avoid damage to any equipment.

You will need

- a level area of ground
- manufacturer's instruction book
- cleaning rags (not cotton waste)
- correct grade/type of oil for engine
- correct grade/type of brake fluid
- correct grades/types of oil/grease for lubrication points
- ready mixed radiator coolant
- tyre gauge (air or air/water as required)
- small tools
- supply of fuel
- clean utensils for fuel and oils
- protective gloves for handling fuel.

You may need

- spare grease nipples
- grease and oil guns
- compressed airline or tyre pump
- spare bulbs and fuses of correct types and sizes
- detergent (to remove grease)
- records and pen/pencil.

Note △ = *important safety point.*

1. General safety

△ Position a tractor on level ground for servicing.

△ Stop tractor engine before doing any servicing operations (except when checking lubrication and cooling systems for leaks after refilling, or when road testing).

△ Remove wristwatch and any rings or bracelets before doing any servicing operations.

△ If servicing a tractor fitted with a loader, lower the loader to the ground if possible, or rest it on a firm support (e.g. a wall) if it must be kept in raised position.

2. Before starting and using tractor

(a) *Checking engine lubrication system*

Tractor *must* be standing level.

Look for any visible oil leaks.

Do not check oil level if engine has been running recently.

Clean around dipstick before pulling it out (to avoid dirt getting in hole).

Inspect dipstick for oil level and for signs of water in the oil (water will give it a frothy brown or creamy appearance and the level on the dipstick will be higher).

(b) *Checking cooling system*

Look for signs of leakage at hoses and connections (presence of water, water marks or stains).

△ Only remove radiator cap if you can bear to keep your hand on the radiator.

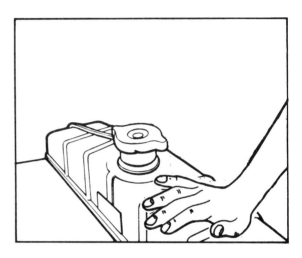

Mix cooling liquid in correct quantities, according to manufacturer's instructions.

Replace dipstick.

Clean around oil filler cap before removing it.

Use a clean can/utensil to add more oil if necessary.

Only add correct grade/type of oil.

After topping up, wipe dipstick clean and re-check oil level.

Take care not to overfill with oil, or damage may be caused.

Replace dipstick.

Wipe off any oil spilled on outside of engine.

Never top up a hot engine with cold cooling liquid.

Top up radiator header tank with ready-mixed cooling liquid to 15mm (½") below filler cap, if necessary (unless a 'no-loss, sealed or semi-sealed system is on the tractor, when manufacturer's instructions must be followed).

Never over-fill the header tank.

Clean radiator core and wire grille (if necessary).

Clean oil cooler core (if fitted).

Record any oil used for topping up (if required).

(c) *Checking air cleaning system*

Clean and service system if required. This will be necessary if the engine has been working in dusty conditions.

Use correct method according to type of cleaner or pre-cleaner (see manufacturer's instructions).

Always use correct type/grade of oil and/or replacement elements.

(d) *Lubricating tractor* (if required — see manufacturer's instruction book)

Always lubricate points in same order.

Always clean each grease nipple and oil filler point before and after lubrication.

Clean grease gun nozzle if necessary before applying to each point.

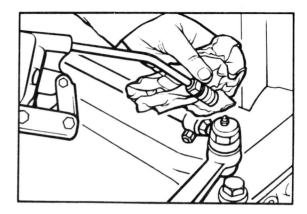

Inject correct amount of grease, — not too little or too much.

Check any grease point which is damaged or does not accept grease and replace it or put right the fault before carrying on.

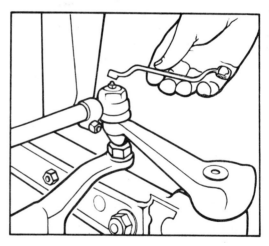

(e) *Checking tyres*

Always check pressures first thing in the morning before using the tractor, (especially if tyres are water-ballasted).

Clean each valve cap before removing it.

Use correct type of gauge (either air, or air/water) to check tyre pressures.

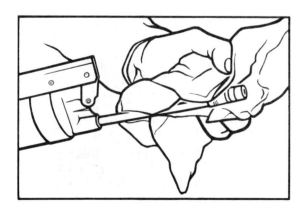

Lubricate each point with correct type/grade of oil/grease according to manufacturer's instructions.

Adjust pressure to suit work to be done, according to manufacturer's instructions.

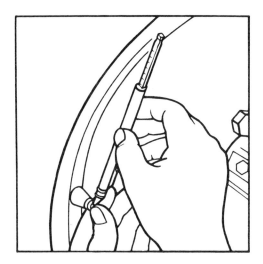

If water ballasted, rotate wheel to place valve at lowest position before checking pressure and/or inflating tyre.

Tyres on same axle must always have equal pressure

Replace and tighten all valve caps, after checking and/or altering pressures and renew any missing caps.

Use detergent to remove all oil or grease from tyre walls and treads.

Remove any embedded nails, stones etc., from tyres

Examine tyre treads and walls for deep cuts and general condition and report if dangerous.

Check and tighten if necessary all wheel and rim bolts (applies particularly to new tractors).

(f) *Checking general condition*

Check level of brake fluid and top up with correct fluid if necessary, according to manufacturer's instructions.

Tighten all slack nuts, bolts and studs.

Check that horn, lights and windscreen wiper will operate.

Replace spent bulbs and blown fuses with new ones of correct rating and type (see manufacturer's instructions.)

Clean all lights and replace any cracked or broken lenses.

△ Replace any cracked or obscured mirror glasses.

△ Check position and security of all guards and safety devices.

△ Check security of safety cab mountings.

△ Clean dirt and grease from all steps, footplates and pedals, with a scraper, or stiff brush and detergent solution, if necessary.

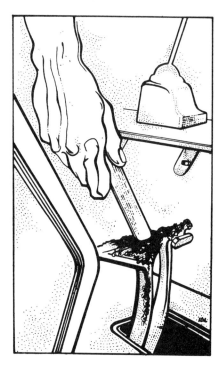

Turn starter switch key to 'ON' and check that all warning lights operate. Switch 'OFF' again.

If required, record tractormeter reading, any items requiring attention at next major service and any items to be reported to employer/manager, (including damage).

3. After using tractor
(Checking fuel system and filling tank)

Always refuel immediately after work, never before work.

△ Always use protective gloves or apply barrier cream when handling fuel and wash hands immediately afterwards.

Clean around fuel cap before removing it.

Place fuel cap on a clean surface after removal.

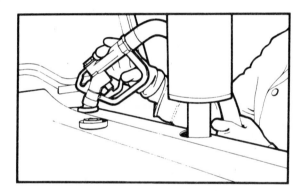

Use clean utensils for refuelling, if from a drum.

Fill tank to within 15mm (½") of the top.

Look for water in fuel sediment bowl, draining it if necessary by slackening screw or removing and emptying bowl.

Replace fuel cap firmly and wipe up any spilled fuel.

Look for fuel leaks and report if necessary.

Record quantity of fuel put in tank (if required).

Tractor servicing detail

Complete this where applicable in the spaces provided, by filling in the information for the tractor which you normally drive.

Tractor - make and model ....

Lubricating oils:

    Engine..

    Transmission ....

    Rear axle and hydraulic system ....

    Steering box (or Power steering)...............................

    Grease..

Air cleaner *(Element no. or oil type)*.........................

Oil filter *(Element no.)* ....

Fuel filters *(Element nos.)* ....

Normal tyre pressures

Front........................ kg./cm^2 (............ lb/sq.in.)

Rear .... kg./cm^2 (............ lb/sq.in.)

You should understand

- the importance of carrying out a daily maintenance plan
- why a fuel tank should be refilled at once after using a tractor, not just before you are ready to use it again
- why a fuel tank must not be filled to the brim.
- the way in which constant lubrication of two surfaces in contact will reduce wear
- the possible results of overfilling the engine's lubrication system, or allowing the oil level to become low
- how cooling systems work
- the importance of the air cleaner and the way it works
- the dangers risked when tyres are neglected
- why you must report any damage immediately to your employer, particularly if a guard has been damaged
- why it is important to keep detailed records of servicing work which you have carried out.

If you do not understand all these points, ask your instructor, manager, employer or course tutor for advice.

Further experience

You will become proficient only by practice. Try to get experience as soon as you can by taking every opportunity offered by your manager or employer to follow up the training you have received.

5 FAMILIARISATION WITH CONTROLS

Your aim

is to be able to:

- recognise each instrument on a tractor and state the meaning and value of the information it gives
- identify and operate each driving control or switch to give a desired result in accordance with the manufacturer's recommendations
- mount and dismount from a tractor in safety
- start a tractor's engine by a safe method and stop it again
- take all precautions to minimise risks of personal injury or injury to other persons or livestock
- take all precautions to avoid causing damage to any equipment or other property.

You will need

- manufacturer's instruction book for the tractor
- ear protectors.

1. General safety

△ Wear suitable solid boots with soles having a good grip.

△ Remove or fasten any loose, flapping clothing e.g. scarf or coat.

△ Fasten back long hair.

△ Do not touch or operate any switch or control on a tractor unless you are sitting in the driving seat.

△ Recognise 'stop' control (fuel cut-off) and be able to use it to stop engine in an emergency.
(Note. The 'stop' control may look different on different tractors and may be operated in different ways — see section 7).

2. Mounting and dismounting

(a) Before mounting

△ Clean anything slippery (e.g. mud) from foot-plate, steps and pedals.

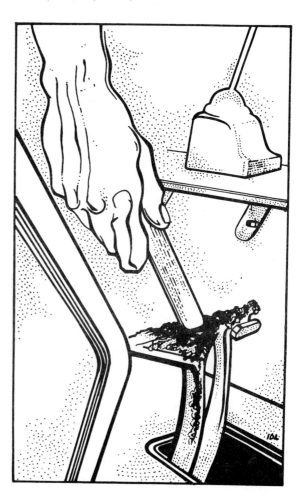

Note △ = important safety point

△ Clean slippery substances from boots.

△ Always mount and dismount from front of rear wheel (where possible).

(c) When dismounting

△ Never try to dismount from a moving tractor.

△ Always stop engine, leaving control in 'stop' position, apply handbrake and leave tractor out of gear.

△ Avoid catching any controls with clothing as you dismount.

△ Remove switch key at end of day.

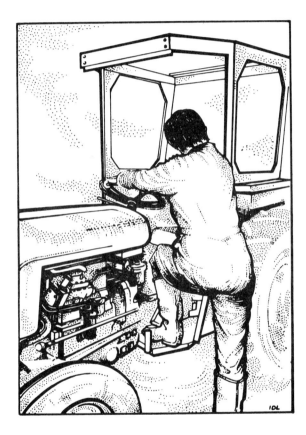

△ Never try to mount a moving tractor.

(b) When mounted

△ Adjust seat so that foot pedals can be fully depressed while you remain seated comfortably.

3. Recognising each instrument

Identify each instrument by its appearance and its symbol.

Understand what each instrument can tell you.

Know the reading which each instrument should indicate when the tractor is operating normally.

Know what to do if an instrument tells you something is wrong or abnormal.

Be able to recognise the following instruments:
A. Tractormeter (revolution counter and hour meter
B. Fuel gauge
C. Ammeter
D. Voltmeter
E. Water temperature gauge
F. Engine oil pressure gauge.

(Some of these instruments may not be fitted to your tractor).

Be able to recognise each warning/indicator lamp by its colour and its symbol and know what it tells you when it is lit.

Complete the diagram on the opposite page as follows:

(i) Draw the positions of all the instruments and warning lamps on the tractor you normally drive.

(ii) Write in the names of each instrument and lamp.

(iii) Alongside the names you have written for each instrument, write the reading you would expect the instrument to show when the engine has been warmed up and is ready to start work.

(If you do not know all the names and readings, look in the Operator's Handbook for the tractor.)

The instruments and warning/indicator lamps of one tractor which you regularly drive

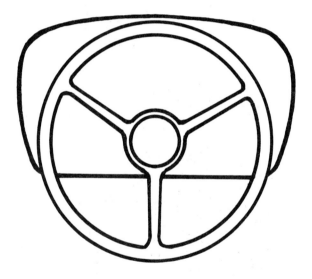

4. Recognising each driving control

Identify each control by its position, appearance and symbol (if applicable)

Understand what will happen when each control is operated.

△ Know the position in which each control must be set before the tractor is started.

Be able to recognise the following controls:
(all these controls may not be found on one tractor)

A. 'Stop' control (fuel cut off)
B. Starter switch key
C. Hand 'throttle' ⎫ Engine
D. Foot 'throttle' ⎭ speed controls
E. Gear shift lever
F. Gear range lever
G. Gear switch/switches
H. Clutch pedal
I. Hand clutch
J. Clutch stop (if fitted)
K. Brake pedals
L. Handbrake (parking brake)
M. Transmission brake
N. Exhaust brake
O. Differential lock pedal or lever.
P. Lighting switch
Q. Direction indicator switch
R. Hazard warning switch
S. Windscreen wiper switch
T. Horn push
U. Fuel tap.
   Any other driving control which is fitted.

Complete the diagram on the opposite page as follows:

(i) Draw the positions of all the controls and switches on the tractor you normally drive.

(ii) Write in the names of each control and switch.

(If you do not know all the positions and names, look in the Operator's Handbook for the tractor).

The driving controls and switches/control levers of one tractor which you regularly drive

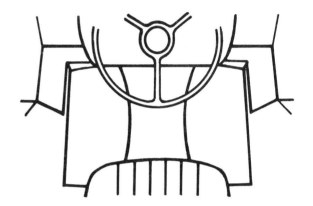

37

5. Preparing to start engine

Carry out morning daily checks and maintenance tasks before starting engine for first time each day.

Open fuel tap (if applicable).

△ Clear any mud/dirt blocking footbrake latch.

Check handbrake is 'on'.

△ Latch footbrakes firmly together.

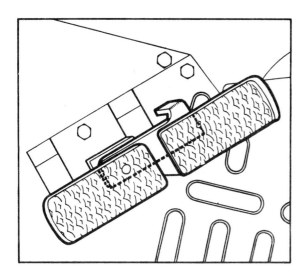

Place 'Stop' control in RUN position.

Open 'throttle' (engine speed control) according to manufacturer's instructions.

△ Set hydraulic controls to LOWER position. ⎫
△ Set p.t.o. in NEUTRAL position. ⎬ if applicable
⎭

△ Do not set these hydraulic and p.t.o. controls in any position other than the starting positions you have been shown, nor touch any of the other controls until you have been instructed in their use.

Place gear levers in correct starting position.

Turn switch key ON

Check that all instruments and warning lights which should work before the engine is started, are working correctly. (Note that some instruments and warning lights will not operate until the engine is running).

6. Starting engine

△ Always sit in the driving seat when starting the engine.

Open 'throttle' (engine speed control).

Operate starter, releasing switch key as soon as engine fires and return 'throttle' to idling position.

Never operate starter continuously for more than about 20 to 25 seconds.

Never operate starter in a series of short bursts.

Always wait until engine and starter are at rest before re-engaging starter.

If engine fails to fire after operating starter for 20 to 25 seconds, wait at least 20 seconds before operating starter again.

Check that all warning lights have gone out when engine is running and that instruments show correct readings. (The readings for a cold engine will differ from those for a hot engine).

Do not allow engine to idle for long periods.

△ Never start or run engine in an enclosed building without adequate ventilation.

Follow cold starting procedure if required, (see manufacturer's instructions for details).

7. Stopping engine

Use 'stop' control to stop engine.

△ Always leave 'stop' control in STOP position after stopping engine.

Turn off switch key after stopping engine.

Remove switch key if tractor will not be used again that day.

Refill fuel tank if tractor will not be used again that day. (See trainee guide M.1.A.1).

You should understand

- why it is important to know what each instrument can tell you and how you should act on the information each instrument supplies.

- how serious damage can be caused if you do not use the correct starting procedure.

- why you are advised to wear ear protectors at all times when driving a tractor.

- that a tractor can be a very dangerous machine and is expensive to buy and repair, so that it must be handled with care and respect at all times.

- why you should never attempt to perform a task, or operate a machine or part of a machine before you have been instructed in that particular activity.

- why you must report any damage, particularly if a guard has been damaged, to your employer.

- why you must at all times comply with legal requirements.

If you do not understand all these points, ask your instructor, manager, employer or course tutor for advice.

Further experience

You will become proficient only by practice. Try to get experience as soon as you can by taking every opportunity offered by your manager or employer to follow up the training you have received.

6 DRIVING STRAIGHT FORWARD AND REVERSE

Your aim

Is to be able to:
— drive a tractor straight forward or in reverse gear
— move smoothly away from rest, forward and in reverse
— stop the tractor smoothly and effectively
— engage and disengage the differential lock correctly
— handle the tractor in accordance with the manufacturer's recommendations
— take all precautions to avoid risks of personal injury or injury to livestock
— take all precautions to avoid causing damage to any equipment or other property.

You will need

— manufacturer's instruction book for the tractor
— ear protectors.

You must

Have received instruction or gained experience in
— familiarisation with controls (trainee guide M.1.A.2).

Note △ = important safety point

1. At all times when driving

Do not rest or 'ride' your foot on the clutch pedal

Only operate clutch at low engine speeds (except at certain times when using p.t.o. driven equipment)

△ Never make unnecessary adjustments to the tractor while it is in motion (e.g. adjusting seat or tools in tool box) — stop before you make the adjustment.

△ It is illegal to carry passengers on the tractor or on linkages or drawbar unless special provision is made.

△ Do not carry on a tractor any tools, items of equipment or guns which may interfere with the tractor controls.

△ Do not drive with a knob attached to the steering wheel rim (it could cause you to break your arm).

△ Keep all footplates, steps and pedals clear of mud, grease, etc and ensure that your boots are clean before mounting the tractor.

2. Moving off

△ Latch brake pedals together.

Move hydraulic controls to TRANSPORT position (if applicable).

△ Depress clutch fully, using ball of foot.

Select a suitable gear for the job, if necessary consulting tractormeter/diagrams/charts.

△ Depress footbrake before releasing handbrake.

Release handbrake fully before moving off.

△ Look around for other traffic, people and children who may be in danger or not know that you are about to move off.

△ Hold steering wheel firmly with thumbs outside the wheel rim alongside your fingers. (Holding the wheel with thumbs inside the rim can lead to dislocated/broken thumbs).

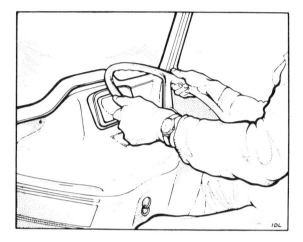

△ Look in direction of travel before tractor begins to move, i.e. forward or backward.

Gently release footbrake whilst releasing clutch.

When clutch has been fully released, remove foot from pedal and increase 'throttle' to obtain desired speed.

△ Never travel at more than a low speed in reverse.

If tractor is fitted with a dual clutch, avoid disengaging p.t.o. when changing gear.

Keep the tractor travelling in a straight line by making small corrections to the steering when required.

Adjust speed with hand throttle; only use foot throttle for road work.

3. Slowing down and stopping

Use 'throttle' lever to slow down, then apply brakes gradually.

△ Do not apply the brakes so hard that the rear wheels skid.

Depress clutch fully when almost stopped, to avoid stalling engine.

When tractor has stopped:
— keep clutch and footbrakes depressed
— move gear levers to neutral
— release clutch
△ — check that differential lock pedal/lever is in DISENGAGED position (see Section four for method of disengaging lock)

— apply handbrake firmly
— release footbrakes
— move hydraulic controls to LOWER position
— stop engine (as in trainee guide M.1.A.2).

4. Using the differential lock

△ Only use the differential lock with a low gear.

△ Never use the differential lock in the higher gears, nor at high speeds or on the highway.

△ Keep the differential lock disengaged unless wheel slip is occurring on one wheel.

Only engage the differential lock when the tractor's rear wheels are turning at the same or almost the same speed, and not whilst wheelspin is occurring.

Release differential lock by pressing the independent brake for the wheel which is tending to slip, or by depressing the clutch.

If you are using the differential lock with the footbrakes latched together (e.g. for trailer work), disengage the lock by steering the tractor first to one side and then the other.

If the differential lock remains in engagement after stopping the tractor, reverse the tractor momentarily to release it.

You should understand

— that a tractor can be a very dangerous machine and is expensive to buy and repair, so that it must be handled with care and respect at all times

— why you should never attempt to perform a task, or operate a machine or part of a machine before you have been instructed in that particular activity

— why you should only use the differential lock at low speeds

— why it is wrong to drive with your foot resting on the clutch

— why brakes must be latched together at all times except when independent braking is essential

— why you must report any damage, particularly if a guard has been damaged, to your employer

— why you must at all times comply with legal requirements.

If you do not understand all these points, ask your instructor, manager, employer or course tutor for advice.

Further experience

You will become proficient only by practice. Try to get experience as soon as you can by taking every opportunity offered by your employer or manager to follow up the training you have received.

7 DRIVING AROUND THE FARM

Your aim

is to be able to:
- steer a tractor around corners in either forward or reverse gears
- use an independent brake where necessary to assist a turn
- drive the tractor in accordance with the manufacturer's recommendations
- take precautions at all times to minimise risks of injury to all persons and livestock
- take precautions to avoid damage to any equipment.

You will need

- manufacturer's instruction book for the tractor
- ear protectors.

Note △ = *important safety point.*

1. General safety

△ Never drive near the edges of ditches or banks, especially if the ground is loose or wet — keep well away.

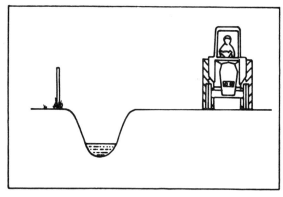

△ Drive slowly on rough ground and when you are turning or are near a ditch.

△ When you are crossing furrows, drive at right angles to the furrows.

△ If the tractor has a safety frame or cab, and it begins to overturn, hold the steering wheel tightly and do not try to leave the cab until the tractor is still.

△ Latch the footbrakes together except when independent brakes are necessary for manoeuvrability.

2. Cornering forwards

Watch out for the rear wheels cutting the corner — steer so that the inside rear wheel travels close to any obstruction but does not touch it.

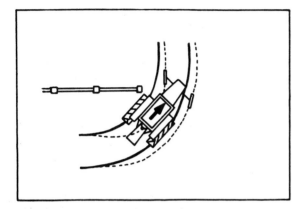

△ Never let the steering wheel spin through your hands when straightening up after making a turn

If a gap is not much wider than the tractor, drive straight into it at right angles until the rear wheels are in the gap, before beginning to turn to one side or the other.

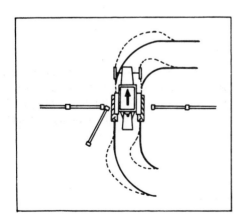

△ Slow down *before* turning a corner — do not make a turn whilst applying the brakes (unless using an independent brake to assist the turn — see section 4).

3. Cornering in reverse

△ Always drive slowly in reverse. Watch out for the front wheels swinging wide — watch for the outside front wheel hitting obstructions.

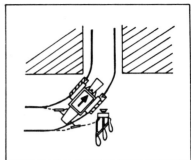

Look over your shoulder on the inside of the corner, giving occasional glances to your outside front wheel for obstructions, if necessary.

△ Never allow the wheel to slip through your hands — it will immediately turn to full lock on one side or the other.

If you are turning into a gap which is not much wider than the tractor, make the turn before you reach the gap and aim to enter the gap with both rear wheels at the same time, i.e. at right angles to the gap.

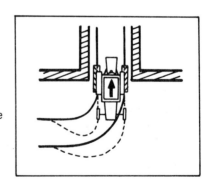

4. Using an independent brake to assist a turn

△ Only use an independent brake at low speeds, never at high speeds.

Use an independent brake to assist a turn only when really necessary.

Always turn the steering wheel to full lock before applying an independent brake to sharpen the turn.

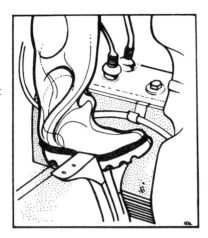

△ Remember that after considerable use of independent brakes it is essential to re-adjust them (see manufacturer's instructions), or wear will make them unbalanced and dangerous when latched together.

5. Driving on sloping ground

△ Never coast the tractor down a hill, in or out of gear.

△ Stop and select a suitable lower gear before climbing or going down a hill — do not try to change gear on the hill.

△ Take particular care in selecting a suitable gear when tractor is fitted with semi-automatic gears.

△ If in doubt about tractor stability when approaching an uphill slope, turn round and climb it in reverse.

△ Never attempt to turn on a steep hillside.

△ Let the clutch engage gradually when starting off on an uphill slope.

△ Brake very gently when reversing down a slope.

△ When driving across a slope, avoid bumps and hollows which may cause the tractor to tilt dangerously. (Using the tractor with the widest permissible track setting will reduce the danger).

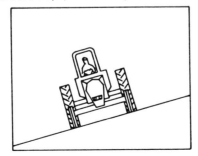

You should understand

- why you must hold the steering wheel firmly, not allowing it to spin through your hands
- why brakes require re-adjustment after considerable independent use
- why it is important to engage a suitable gear before climbing or going down a hill
- that a tractor can be a very dangerous machine and is expensive to buy and repair, so that it must be handled with care and respect at all times
- that you should never attempt to perform a task, or operate a machine or part of a machine before you have been instructed in that particular activity
- that you must report any damage, particularly if a guard has been damaged, to your employer
- that you must at all times comply with legal requirements.

If you do not understand all these points, ask your instructor, manager, employer or course tutor for advice.

Further experience

You will become proficient only by practice. Try to get experience as soon as you can by taking every opportunity offered by your manager or employer to follow up the training you have received.

8 ATTACHING MOUNTED IMPLEMENTS TO THE 3-POINT LINKAGE

Your aim

is to be able to:

— prepare the tractor linkage and drawbar (if applicable) for attachment
— attach the machine or implement to the linkage by the correct method
— fit any stabilising bars necessary
— remove the machine from the tractor and store it securely
— take all precautions to minimise risks of personal injury or injury to other persons or livestock
— take all precautions to avoid causing damage to any equipment or buildings.

You will need

— suitable rear-mounted machine
— top link
— wooden blocks and stands to support detached machine
— any tools necessary
— manufacturers' instruction books for the tractor and for the machine.

You may need

— alternative lower link ball ends
— stabiliser (to suit the machine being attached)
— spare linchpins or linchpin spring rings.

You must

have received instruction or gained experience in:
— driving around the holding (trainee guide M.1.A.4).

Note

1. Although this trainee guide contains a brief outline of ways of adjusting wheel adhesion and tractor stability in section 2 (b), it is unlikely that you will be instructed in this until later. The outline has been included in this guide to indicate the stage of the work at which these adjustments would be made if it were necessary for the correct use of the machine/implement.

2. The guide also indicates the correct points in the sequence when a p.t.o. shaft and hydraulic couplings would be connected and disconnected. Normally, you would be instructed in these activities at a later date. Trainee guides M.1.B.3 and M.1.B.4., together with M.1.C.1. to M.1.C.6. give further details of all these activities

Note △ = important safety point.

1. General safety

△ Do not wear loose clothing while working on a tractor, especially when using a p.t.o. driven machine.

△ Move p.t.o. lever to NEUTRAL, apply the parking brake and stop engine before dismounting, for any reason, from a tractor upon which a machine is mounted.

△ Know and observe the regulations regarding the age of a person permitted to remove a guard from a p.t.o.

△ Know and observe the regulations relating to safety devices and to the conditions under which a tractor must be fitted with an approved safety frame/cab.

2. Preparing tractor

(a) Adjust linkage and drawbar.

Position lower links correctly or remove lower link ball ends and replace with alternative lower link ball ends to suit category of machine to be mounted (see manufacturer's instructions).

Adjust check chains if necessary to allow 25 mm. (1") total float when lower links are fully raised.

If the lower links have angled ends, check that the ends are parallel to each other.

Ensure that the linkage is completely free.

△ Set left link to correct length (see manufacturer's instructions).

△ Ensure that all linchpins are fitted and fully in position.

△ Check all linchpin spring rings for good condition (snapping into position) and replace with new spring rings where necessary.

Position and firmly secure drawbar so as not to foul machine when it is attached.

Check that there will be sufficient clearance between the machine and rear of tractor.

Remove p.t.o. cover and stow away safely (if applicable).

(b) Adjust wheel adhesion and tractor stability (if necessary).

Refer to manufacturer's instruction book.

Jack up tractor using a safe method.

Alter wheel track width and/or attach wheel weights, frame weights, cage wheels and strakes (as required).

Ballast tyres (if required).

Adjust ground clearance (if applicable).

(c) Check and if necessary adjust tyre pressures.

Refer to manufacturer's instruction books for correct pressures.

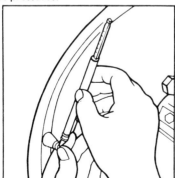

Use air/water pressure gauge for ballasted tyres.

Check and adjust pressure of a ballasted tyre with valve at bottom of wheel.

△ Do not over-inflate a ballasted tyre.

Do not reduce tyre pressure below the recommended minimum.

3. Preparing machine (if applicable)

Fit machine linkage points for lower links into correct position for category of machine or work which is to be done (see manufacturer's instructions).

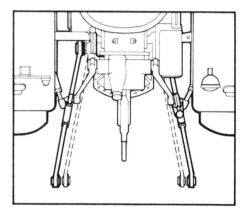

4. Attaching the machine

(a) Reverse tractor to machine.

Reverse slowly, at right angles to linkage points of machine.

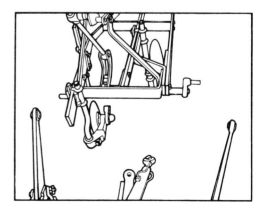

Line up linkage and linkage points while reversing.

Use height/position control to set left link at correct level.

Position top link so that it will not be in the way.

Position p.t.o. shaft out of the way of the linkage and the stowed drawbar (if applicable).

To make small, careful tractor movements, use a low gear and low engine speed, and slip the clutch to control the speed and distance of the movements. (Avoid slipping the clutch too much; damage will be caused).

While reversing, hold right link as far out as check chain allows and position tractor so that right link ball end is just clear of the end of the right linkage hitch point. (It does not matter if the right link ball end is not at the correct height).

△ Apply handbrake and move all controls to NEUTRAL before dismounting from tractor.

(b) Attach links

△ Do not use a finger to line up any holes

Attach left link and secure it with its linchpin.

Always fit linchpin and secure it with the spring ring as soon as each link is in position.

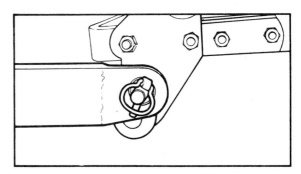

Use levelling handle and/or machine crossbar adjustment to line up right link. (The machine may also have to be lifted slightly on the left link, using the height control, to line up the right link).

△ If use of height control is necessary, tell other persons to stand well clear, and then operate control whilst sitting on tractor seat.

Attach right link and secure with linchpin.

Check top link to ensure that it is correct for the machine.

Set turn buckle so that it is exactly in centre of top link.

Attach top link to tractor (in correct hole for work to be done).

Adjust top link length for attachment by holding outer end still and turning turn buckle.

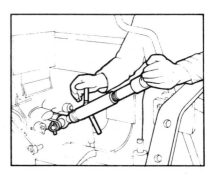

Attach top link to head stock of machine, in correct hole for work to be done (see manufacturer's instructions).

Always attach links in correct order, i.e. left, right and finally the top link.

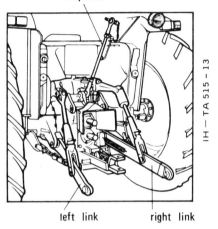

Top link

left link   right link

△ Always use the linkage and machine adjustments to line up links — do not try to push or lift the machine.

△ Do not try to move a machine by lifting it by its soil working parts.

△ Sit in driving seat before operating any of the tractor controls.

(c) Level machine and fit stabiliser bars/chains (if necessary).

See manufacturer's instructions.

Fully secure any linchpins.

(d) Connect p.t.o. drive and any hydraulic connections to external services (if applicable).

(e) Prepare to drive tractor and machine away.
Raise machine on linkage.
Remove any stands, blocks or chocks which have been supporting machine.
Ensure that tank contains sufficient fuel for planned work.

5. Parking tractor and machine temporarily
(with engine stopped)

△ If machine cannot be locked up on a lift latch, lower it to ground whenever parking tractor.

△ Before doing any work beneath a raised machine, support it on wooden blocks or a stand in addition to any lift latch which may be in use.

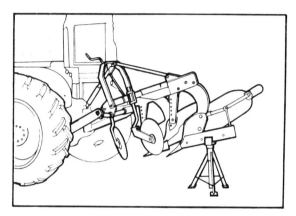

△ Do not support machine on stones, bricks or steel (other than an axle stand of suitable capacity).

6. Detaching machine from tractor

(a) Lowering machine

Before lowering machine, set rate of drop control according to weight of machine (if necessary).

△ Whenever possible, ensure that tractor and machine are on level ground.

Drive tractor into required position and place any necessary wooden blocks and/or stands in correct positions to support machine before lowering machine.

△ After lowering, check that machine is fully supported and secure before detaching it from tractor.

(b) Detaching machine

△ Stop engine, apply parking brake and check that gear and p.t.o. levers are in NEUTRAL positions.

△ Do not stand between tractor and machine.

Follow correct order of working.

(i) Disconnect all hydraulic connections, (if applicable)

(ii) Disconnect p.t.o. (if applicable)

△ Replace p.t.o. cover over p.t.o. as soon as shaft has been disconnected.

Do not engage p.t.o. drive after shaft has been disconnected.

(iii) Disconnect three point linkage.

Remove stabiliser (if fitted).

Disconnect linkage in order — top link, offside link and lastly nearside link.

Stow all linchpins safely.

(c) Driving tractor away.

Before moving away, ensure that no remaining connection exists between tractor and machine.

You should understand

— why it is important that the three point linkage is always connected and disconnected in the same order.

— why it is important that the right link should not be allowed to become screwed too far out.

— the importance of immunisation against tetanus.

If you do not understand all these points ask your instructor, manager, employer or course tutor for advice.

Further experience

You will become proficient only by practice. Try to get experience as soon as you can by taking every opportunity offered by your employer or manager to follow up the training you have received.

9 HITCHING TRAILED MACHINES

Your aim

Is to be able to:
- attach and detach a pick-up hitch, a linkage drawbar or a swinging drawbar to and from a tractor and prepare to couple a trailed machine to the hitch gear
- couple a trailed machine to and uncouple it from a tractor
- take all precautions to minimise risks of personal injury or injury to other persons or livestock
- take all precautions to avoid causing damage to any equipment.

You will need

- manufacturer's instruction book for the tractor
- hitch gear as required
- manufacturer's instruction book for the machine
- tools as required (e.g. for the removal of clevis adaptors).

You must

Have received instruction or gained experience in:
- familiarisation with controls (trainee guide M.1.A.2)
- driving straight forward and reverse (trainee guide M.1.A.3).

You may also need

(According to the type of machine to be hitched to the tractor) to have received instruction or gained experience in:
- driving around the holding (trainee guide M.1.A.4).

Note △ = important safety point

1. General safety

△ Do not touch or operate any switch or control on a tractor unless you are sitting in the driving seat.

△ Move p.t.o. lever to NEUTRAL, apply the parking brake and stop engine before dismounting for any reason from a tractor to which is coupled a p.t.o. driven trailed machine.

△ Do not allow children to travel on a tractor or trailed machine.

△ Know and observe the regulations regarding the age of a person permitted to remove a guard from a p.t.o.

△ Know and observe the regulations relating to safety devices and to the conditions under which a tractor must be fitted with an approved frame/cab.

△ Always ensure that tractors are big enough and have sufficient braking power to be safe, when machines are attached. Especially when:
- the machine can be loaded
- there is sloping land.

2. Adjust wheel adhesion and tractor stability (as necessary)

Refer to manufacturer's instruction book.

(a) Jack up tractor
See trainee guide M.1.C.1 — Jacking up a tractor.

(b) Alter wheeltrack width
See trainee guide M.1.C.4 — Altering wheel track widths.

(c) Fit wheel and/or frame weights
See trainee guide M.1.C.2 — Fitting wheel and frame weights.

(d) Fit cage wheels or strakes
See trainee guide M.1.C.3 — Fitting cage wheels and strakes.

(e) Adjust ground clearance (if applicable)
Refer to manufacturer's instruction book.

(f) Ballast tyres
See trainee guide M.1.C.5 — Water ballasting tyres.

(g) Check and adjust tyre pressures (if necessary).

Refer to manufacturer's instruction book.

Use an air/water pressure gauge to check ballasted tyres.

Check pressure of a ballasted tyre with valve at the bottom of wheel.

△ Do not over-inflate a ballasted tyre.

Inflate a ballasted tyre with the valve at the bottom of the wheel.

Do not reduce tyre pressure below the recommended minimum.

Drive tractor to place point of hitch under centre of drawbar ring — allow for difficulty in seeing positions of hitch and ring from driver's seat.

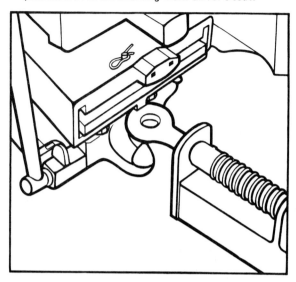

3. Pick-up hitch — Attachment and coupling

(a) Attaching a pick-up hitch.

See manufacturer's instruction book.

Remove linkage drawbar and/or top link from tractor (if attached) and store in correct place.

Remove swinging drawbar if attached and store in correct place (unless drawbar and pick-up hitch are combined in one attachment which can be reversed).

Securely attach all components of pick-up hitch.

(b) Coupling a tractor with a pick-up hitch to a machine.

Ensure that hitch hydraulics are operating.

Reverse tractor to machine at a speed allowing complete control over tractor while lining up with machine.

Do not lower hitch so far that it touches the ground.

Use correct hydraulic control to lift machine to towing position.

△ Operate all tractor controls only from driver's seat.

Select a suitable engine speed.

Either — secure raised hitch with latching device before lowering hydraulic lever
or — isolate hitch from hydraulics.

Do not allow hitch to be held up by hydraulic pump.

4. Attaching a linkage drawbar

See manufacturer's instruction book.

Remove top link from tractor (if attached) and store in correct place.

Either — remove pick-up hitch (if attached) and store in correct place
or — isolate pick-up hitch from hydraulics.

Alter or adjust lower link ball ends as necessary to give correct category.

Attach linkage drawbar.

Securely attach any necessary stay-bars or stabilisers.

Check condition of spring rings on linchpins and replace if they do not snap into position.

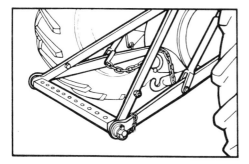

If trailed machine is in store at a distance from the tractor, ensure that correct types of drawbar pin and linchpin are safely stowed on tractor.

5. Attaching a swinging drawbar

See manufacturer's instruction book.

Remove linkage drawbar and top link (if either is attached) and store in correct place.

Remove pick-up hitch if attached and store in correct place (unless drawbar and pick-up hitch are combined in one attachment which can be reversed).

Locate drawbar in central position and attach securely.

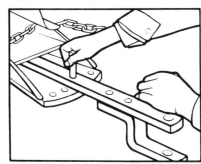

If trailed machine is in store at a distance from the tractor, ensure that correct type of drawbar pin and linchpin are safely stowed on tractor.

6. Coupling a tractor with a linkage drawbar or swinging drawbar to a machine

Reverse tractor to machine at a speed allowing complete control over tractor while lining up with machine.

△ Apply parking brake and move all controls to neutral before dismounting from tractor.

△ Line up drawbars by manoeuvring tractor and/or by adjusting drawbar jack of machine — do not manhandle a heavy machine.

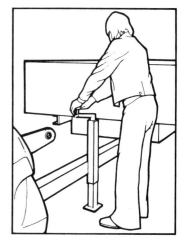

△ When adjusting drawbar jack, stand with feet away from the area of ground beneath the drawbar.

△ Ensure that only a single hitch point is coupled between the two tongues of a clevis — other combinations of hitch points are dangerous, especially two double clevis hitches.

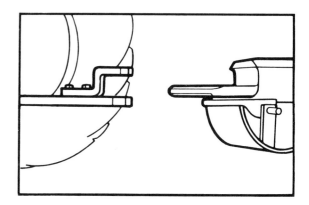

△ If using a linkage drawbar
 - avoid using a high hitch point
 - isolate linkage from hydraulics.

If coupling to a p.t.o. driven machine
 - remove any clevis adaptation which may foul p.t.o. shaft guard.

Couple an offset machine in transport position unless beginning work immediately.

Keep all drawbar jacks well greased.

△ Fit drawbar pin while standing next to drawbar — not by reaching down from driver's seat.

△ Do not attempt to tow a machine from any point on a tractor except from a drawbar.

△ Use the correct type and size of drawbar pin.
If coupling to a p.t.o. driven machine
 - do not use a drawbar pin which might foul the p.t.o. shaft guard (e.g. a pin with a handle).

△ Secure drawbar pin at its lower end with a linchpin.

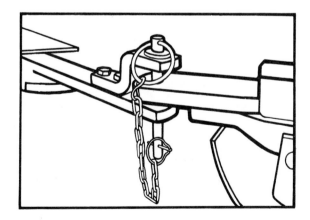

△ Check condition of spring ring on linchpin and replace if it does not snap into position.

Remove any chocks, stands or blocks which have been supporting and/or securing machine.

7. Preparing to drive away

Connect hydraulic hose(s) to tractor (if applicable) — see trainee guide M.1.B.4 Connecting machines to the tractor's hydraulic services.

Fix drawbar jack securely in transport position, well clear of ground.

Remove any chocks from in front of machine wheels.

Check that machine is correctly prepared for transport.

Raise link arms, if not in use, to avoid fouling machine drawbar, hydraulic hose(s) etc.

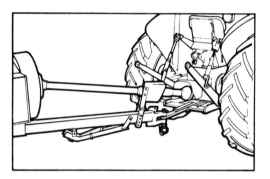

Ensure that fuel tank contains sufficient fuel for planned work.

If coupling to p.t.o. driven machine, ensure that p.t.o. shaft is
either — anchored securely to transport bracket.
or   — correctly connected to p.t.o. (see trainee guide M1.B.3. Attaching p.t.o. shafts).

8. Uncoupling a trailed machine

Disconnect p.t.o. drive (if applicable) before transporting machine/ to storage place — see trainee guide M.1.B.3. Attaching p.t.o. shafts.

△ Apply parking brake and move all controls to a neutral position before dismounting from tractor. Place chocks behind and in front of each machine wheel.

Lower drawbar jack and secure.

△ Set external hydraulic control valve in neutral position before disconnecting hydraulic hoses, if applicable.

Follow key points for stowing hoses and protecting hydraulic connections — see trainee guide M.1.B.4, Connecting machines to the tractor's external hydraulic services.

Remove drawbar pin and store in correct position.

Disconnect any operating wires or cords.

9. Uncoupling a pick-up hitch

△ Operate tractor controls only from the driver's seat.

Select suitable engine speed.

Activate hitch hydraulics (if isolated).

Use correct hydraulic control to lift hitch in order to release latching device.

After releasing latching device, lower hitch to release drawbar ring.

Check that there is no further contact between tractor and machine and move tractor forwards, clear of machine.

Stop when clear of machine and:
either — raise hitch, secure with latching device and lower hydraulic lever
or — raise hitch and isolate from hydraulics.

10. Detaching hitch gear from tractor

See manufacturer's instruction book where necessary.

Store all parts together in the correct storage place.

Stow linchpins and locating pins safely if they are not to be used again immediately.

You should understand

— why a pick-up hitch should be held in the raised position on a latching device and not on the hydraulic pump

— why the spring rings of all linchpins must often be checked for good condition

— why a coupling between two double clevis hitches can be dangerous.

— why it is dangerous to hitch a trailed machine to a high hitch point.

— why it is dangerous to use an incorrect drawbar pin or a pin which is unsecured at its lower end.

If you do not understand all these points ask your instructor, manager, employer or course tutor for advice.

Further experience

You will become proficient only by practice. Try to get experience as soon as you can by taking every opportunity offered by your employer or manager to follow up the training you have received.

10 ATTACHING PTO SHAFTS

Your aim

is to be able to:

- connect the p.t.o. shaft of a p.t.o. driven machine to a tractor's power take off
- disconnect the p.t.o. shaft from the tractor
- take all precautions to minimise risks of personal injury or injury to other persons or livestock
- take all precautions to avoid causing damage to any equipment.

You will need

- manufacturer's instruction book for the tractor
- manufacturer's instruction book for the machine which will be driven through the p.t.o. shaft
- tools as required.

You must

have received instruction or gained experience in:

- attaching mounted machines to the three-point linkage (trainee guide M.1.B.1), or
- hitching trailed machines (trainee guide M.1.B.2).

△ never approach a p.t.o shaft while the tractor engine is running.

1. Attaching or coupling machine to tractor

As appropriate, see

- Attaching mounted machines to the three point linkage (trainee guide M.1.B.1)
- Hitching trailed machines (trainee guide M.1.B.2).

△ Know and observe the regulations regarding the age of a person permitted to remove a guard from a p.t.o.

△ Ensure that all regulations relating to the guarding of p.t.o. shafts are complied with at all times.

△ Remove or fasten loose clothing; fasten back long hair, scarves etc.; roll up shirt or overall sleeves.

2. Moving machine into operating position

△ Apply parking brake, move all controls to a neutral position and stop engine before dismounting from tractor.

Disconnect p.t.o. shaft from transport bracket (if applicable) before moving machine to operating position.

If applicable, move machine drawbar from transport position to operating position (see manufacturer's instruction book).

If machine is mounted, adjust levelling handle and top link until machine is level.

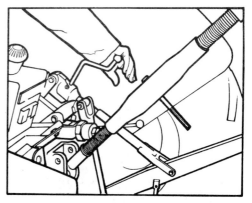

Note △ = *important safety point.*

3. Connecting p.t.o. shaft to tractor p.t.o.

If tractor has a high speed and a low speed p.t.o., use correct p.t.o. to suit the machine to be operated (see machine manufacturer's instructions).

Remove p.t.o. cover and stow away safely.

Check the p.t.o. shaft insertion overlap; there must be 150mm (6") minimum engagement at maximum extension and at least 20mm (½") further possible engagement at maximum telescoping (see manufacturer's instructions).

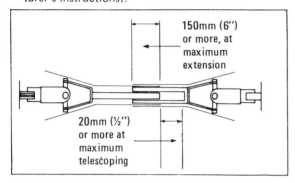

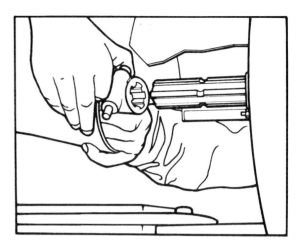

The two halves of the p.t.o. shaft must have the correct length and section size for the machine in use and must be securely connected with the universal joints in the same plane. (see manufacturer's instructions).

Inspect p.t.o. and shaft splines for freedom from damage and foreign matter.

△ Check that p.t.o. shield is securely attached to the tractor and undamaged.

△ Clean inside of p.t.o. shaft guard, (if necessary).

Lubricate well the p.t.o., p.t.o. shaft, plungers and guard bearings with the correct lubricants, if required. (see manufacturer's instructions).

Check p.t.o. shaft for free telescoping action.

Inspect plungers for wear, and replace if worn.

Rotate p.t.o. by hand, if necessary to line up splines on p.t.o. and p.t.o. shaft.

Squeeze plungers and slide universal joint on to p.t.o. until it locks securely into locating groove.

Do not hammer the universal joint on to the p.t.o.

Check that guard tubes are correctly and securely seated (and anchored by chain, if fitted).

Check that p.t.o. shaft does not foul drawbar pin — change pin for different type (e.g. flat-headed) if necessary.

4. Manoeuvring a tractor and machine with p.t.o. engaged or disengaged

Take particular care when manoeuvring a tractor and a mounted or trailed p.t.o. driven machine (see trainee guide M.1.B.5 — Manoeuvring with a mounted machine, or M.1.B.6 — Manoeuvring with a two wheeled trailer or machine, as appropriate).

If tractor has a dual clutch, avoid disengaging p.t.o. when changing gear.

5. Disconnecting p.t.o. shaft from tractor p.t.o.

△ Apply parking brake, disengage p.t.o., move all controls to a neutral position and **STOP ENGINE** before attempting to disconnect the p.t.o shaft.

Squeeze plungers and slide universal joint off p.t.o.

Do not hammer a universal joint when removing it from the splined shaft.

△ Do not leave any part of a p.t.o. shaft attached to tractor when you have finished with it.

△ Replace p.t.o. cover over p.t.o. as soon as shaft has been disconnected.

If necessary, change drawbar to transport position *after* disconnecting p.t.o. shaft.

Do not engage p.t.o. drive after shaft has been disconnected.

6. Remove or uncouple machine

See trainee guide M.1.B.1 (for mounted machines) or M.1.B.2 (for trailed machines) as appropriate.

You should understand

— why it is necessary to follow the regulations relating to the guarding of p.t.o. shafts

— why it is important that each machine is only connected to whichever p.t.o. shaft (high speed or low speed) is correct for that machine

— why a p.t.o. shaft should always be checked before use to ensure that it has the correct insertion overlap.

— why the tractor engine must be stopped before attempting any work on the p.t.o. shaft.

If you do not understand all these points, ask your instructor, manager, employer or course tutor for advice.

Further experience

You will become proficient only by practice. Try to get experience as soon as you can by taking every opportunity offered by your manager or employer to follow up the training you have received.

11 CONNECTING MACHINES TO THE EXTERNAL HYDRAULIC SERVICES

Your aim

Is to be able to:

— connect and disconnect hydraulic connections between a trailed or mounted machine and the external hydraulic services of a tractor

— carry out the connection or disconnection in accordance with the manufacturer's recommendations

— take precautions to minimise risks of injury to all persons and livestock

— take precautions to avoid damage to any equipment.

You will need

— manufacturer's instruction book for the tractor

— manufacturer's instruction book for the machine which is to be powered or controlled through the tractor's external hydraulic services

— tools as required.

— first aid equipment which complies with the Health & Safety (First Aid) Regulations, 1981.

You must

Have received instruction or gained experience in:

— attaching mounted machines to the three-point linkage (trainee guide M 1.B.1)

or

— hitching trailed machines (trainee guide M 1.B.2) as appropriate.

Note △ = important safety point.

1. Attaching/coupling machine to tractor

As appropriate, see trainee guides:

— Attaching mounted machines to the three-point linkage (trainee guide M 1.B.1)

— Hitching trailed machines (trainee guide M 1.B.2).

2. Connecting hydraulic pipes

△ Always clean oil off your skin as quickly as possible to avoid risk of dermatitis.
See manufacturer's instruction books to check whether it is necessary to add extra oil to the tractor's transmission and how much extra oil is needed or may be added.

△ Stop engine and release pressure in hydraulic system by setting external hydraulic controls in neutral position.

Check that hoses are undamaged — report any damage to employer/supervisor.

Remove caps/dust covers and stow away in correct positions.

Check that connections are absolutely clean before connecting.

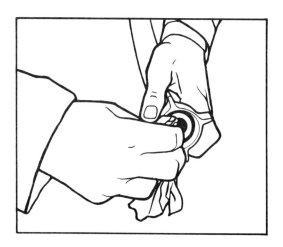

Ensure that connections are fully engaged or screwed in firmly but are not over-tight.

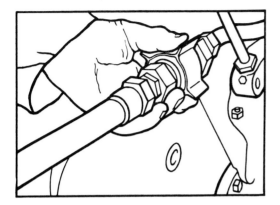

△ Do not use more than hand force on a screw-type coupling.

Check hoses are:
— long enough to allow necessary movements
— not twisted
— secured by correct brackets, hooks etc., to avoid rubbing on any moving parts of machine.

△ Do not secure hydraulic hoses with string or wire.

Set external hydraulic controls in live position.

Check hydraulic system for visible oil leaks.

△ Stop work and inform employer/supervisor immediately if leaks are found.

△ Do not attempt to work on hydraulic systems which are under pressure.

△ Release pressure in hydraulic system before tightening couplings or seals.

△ If a hydraulic hose should burst, do not attempt to approach or handle it without first relieving the pressure.

3. Disconnecting hydraulic pipes

△ Stop engine and set external hydraulic controls in neutral position to release pressure in system — do not attempt to disconnect a hose which is under pressure.

Disconnect each hose in turn and stow it in correct position.

Ensure that connections and dust caps are absolutely clean.

Replace dust caps securely as soon as each hose is disconnected.

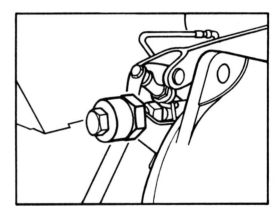

Do not overtighten hoses and dust caps.

Ensure that hoses are not kinked or twisted when stowed.

Store rams in retracted position when possible.

Grease exposed shafts of rams before storing.

3. Disconnecting hydraulic pipes

△ Stop engine and set external hydraulic controls in neutral position to release pressure in system — do not attempt to disconnect a hose which is under pressure.

Disconnect each hose in turn and stow it in correct position.

Ensure that connections and dust caps are absolutely clean.

Replace dust caps securely as soon as each hose is disconnected.

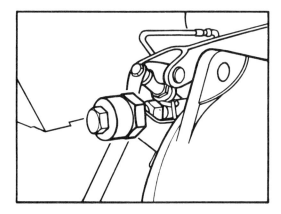

Do not overtighten hoses and dust caps.

Ensure that hoses are not kinked or twisted when stowed.

Store rams in retracted position when possible.

Grease exposed shafts of rams before storing.

12 MANOEUVRING WITH MOUNTED MACHINES

Your aim

is to be able to drive a tractor and mounted machine, implement in forward and reverse gears so that:

— no damage is caused to any equipment or buildings

— risks of personal injury or of injury to other persons or livestock are minimised

You will need

Manufacturer's instruction book for the mounted machine (if applicable).

You should

have received instruction or gained experience in:

— attaching mounted machines (trainee guide M.1.B.1).

1. General safety

△ Move p.t.o. lever to neutral, apply the parking brake and stop engine before dismounting for any reason from a tractor upon which a machine is mounted.

△ Do not allow children to travel on a mounted machine.

△ Do not allow passengers to travel on a tractor, or on any machine which is not fitted with an operator's platform or seat.

2. Manœuvring a tractor and mounted machine

△ Latch the brakes together.

When turning make allowances for the machine, which will need more clearance to swing on the outside of the turn — a long or wide machine needing most clearance.

Choose the correct gear for the speed required — do not use the clutch to control the speed.

Note: △ = *important safety point.*

△ Drive slowly on corners, particularly when a heavy machine is mounted on the tractor.

△ Drive slowly in reverse.

Always move p.t.o. lever to neutral before reversing.

Drive slowly over rough ground, particularly if the machine is heavy and/or is supported on the linkage by a **lift latch**.

Move p.t.o. lever to neutral when not in work.

△ When parking tractor, lower machine to ground if not locked up on a lift latch.

You should understand

- why it is important to move the p.t.o. lever to neutral before reversing

- the responsibilities of a worker under the Agriculture (Field Machinery) Regulations 1962.

If you do not understand all these points, ask your instructor, manager, employer or course tutor for advice.

Further experience

You will become proficient only by practice. Try to get experience as soon as you can by taking every opportunity offered by your employer or manager to follow up the training you have received.

13 MANOEUVRING WITH A 2-WHEELED TRAILER OR TRAILED MACHINE

Your aim

Is to be able to:

— manoeuvre a tractor and two-wheeled trailer/trailed machine, in forward and reverse gears, under easy and difficult conditions i.e.
  involving straight runs and corners,
  level and sloping surfaces,
  smooth and rough surfaces,
  open spaces and confined areas.
— manoeuvre a tractor and two-wheeled trailed p.t.o.-driven machine.

You will need

Manufacturer's instruction book for the trailed machine (if applicable).

You must

have received instruction or gained experience in:

— hitching trailed machines (trainee guide M.1.B.2)
— attaching p.t.o. shafts (trainee guide M.1.B.3)
— connecting machines to the tractor's external hydraulic services (trainee guide M.1.B.4).

1. Coupling trailer/trailed machine to tractor

Refer to manufacturer's instruction book (if applicable).

Refer, if necessary, to:
— hitching trailed machines (trainee guide M.1.B.2)
— attaching p.t.o. shafts (trainee guide M.1.B.3)
— connecting machines to tractor's external hydraulic services (trainee guide M.1.B.4).

2. Preparing to drive away

Refer, if necessary, to:
— Hitching trailed machines (trainee guide M.1.B.2, Section 7)

3. At all times when manoeuvring with a tractor and trailer/trailed machine

△ Do not allow passengers to travel on the drawbar of a trailer/trailed machine or on any trailed machine which does not have a special place for carrying an operator.

△ Do not allow children to ride on any trailed machine.

△ Do not allow children to travel on a trailer unless the trailer has four sides which all extend above the height of the load.

△ Ensure that any passengers on an empty trailer are seated in the centre of the trailer.

△ If another operator is riding on a trailed machine (e.g. a drill) ensure that he is safely in position and correctly guarded before moving off.

△ Ensure that no children or other people are in possibly dangerous positions near the tractor and trailer/machine when moving off.

Note △ = important safety point.

△ Latch independent brakes together and keep them coupled.

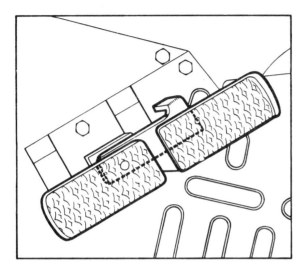

△ Give a wide clearance to rivers, ditches, banks and quarries, etc.

△ Do not attempt to turn on a steep hillside.

4. Moving away from rest

Allow for weight of trailed machine or trailer and any load when selecting gear and engine speed.

Engage clutch gradually.

Do not use the clutch to control the speed of the tractor and trailer or machine.

5. Driving forwards

△ Always drive at a speed which allows complete control.

△ Slow down when crossing uneven ground.

△ Drive across furrows at right angles to the line of the furrows.

△ Do not tow a heavily loaded trailer or a heavy machine across the slope of a steep hillside.

△ Do not attempt to turn on a steep hillside.

△ Stop and select a suitable gear before climbing or going down a steep hill when towing a loaded trailer or heavy machine — do not try to change gear on the hill.

△ Fit sufficient front frame weights or front wheel weights to the tractor before climbing a steep hill towing a loaded trailer or heavy machine.

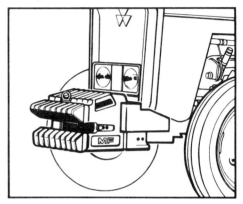

△ Ensure that differential lock is disengaged before making turns or driving on to the highway.

△ Slow down before cornering — ensure that speed is sufficiently slow to avoid having to apply the brakes whilst making the turn.

When turning, do not turn so tightly that the tractor's rear wheels foul the drawbar, p.t.o. shaft or hydraulic hoses of a trailer or machine.

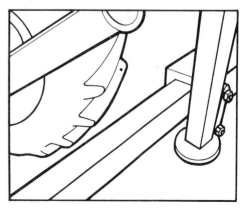

Allow for width of trailer/trailed machine when manoeuvring in confined spaces, through gateways and when passing any other vehicles.

Disengage the p.t.o. drive before turning a sharp corner with a trailed machine (unless the machine has been designed to turn sharply — see manufacturer's instructions).

When using a pressure hitch:
— know how to operate hydraulic controls
— do not apply more pressure than the conditions require.

6. Driving in reverse with a trailer or trailed machine

△ Always drive slowly in reverse gears.

Disengage p.t.o. drive before reversing.

Know in which direction to steer in order to make the trailer or machine follow the desired course.

Do not allow the tractor's rear wheels to foul the drawbar, p.t.o. shaft or hydraulic hoses of the trailer or machine.

△ Do not attempt to turn on a steep hillside.

You should understand

— why it is important to follow the regulations relating to the carrying of passengers on trailers and trailed machines.
— why the independent brakes must be kept latched together at all times when towing a trailer or trailed machine.
— why it is dangerous to:
    turn on a steep hillside,
    drive across a steep hillside, particularly with a heavy load,
    change gear on a hill,
    tow a heavy load up a steep hill without having enough weight on the front of the tractor.
— why it is dangerous to brake during a turn when towing a trailer or machine
— why the p.t.o. must be disengaged before reversing
— relevant safety legislation.

If you do not understand all these points, ask your instructor, manager, employer or course tutor for advice.

Further experience

You will become proficient only by practice. Try to get experience as soon as you can by taking every opportunity offered by your employer or manager to follow up the training you have received.

14 JACKING UP TRACTORS AND WHEELED MACHINES

Your aim

Is to be able to:
- jack up the vehicle in preparation for carrying out maintenance or other activities
- jack up the vehicle in accordance with the manufacturer's recommendations
- take all precautions to minimise risks of personal injury or injury to other persons or livestock
- take all precautions to avoid causing damage to any equipment.

You will need

- a firm, level site on which to jack up the vehicle
- manufacturer's instruction book for the vehicle
- jack of suitable capacity — (preferably a trolley jack)
- axle stands having a suitable capacity, or suitable wooden blocks to support the necessary weight
- 6 wheel chocks
- any tools necessary.

You may need

A wide, flat board or metal plate for each jack and axle stand/block, (if the site surface is not absolutely hard).

You must

have received instruction or gained experience in:
- driving around the holding (trainee guide M.1.A.4).

Note △ = important safety point.

1. General safety

△ Always select a firm, level site on which to jack up a vehicle.

△ Stop engine before jacking up vehicle.

△ Always remove wristwatch and any rings or bracelets before doing any servicing operations.

2. Jacking up the vehicle

Disconnect any trailer or implement (see trainee guides M.1.B.1 to M.1.B.4 as appropriate, for more details).

△ Before jacking, place chocks tightly in front of and behind each wheel which will stay on the ground, then slacken the nuts on the wheel to be removed by half a turn.

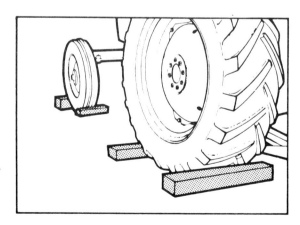

△ Take extra care to be sure that chocks cannot slip if using a screw or bottle jack.

△ Always use axle stands or metal/wooden blocks as well as jacks when you are working on a jacked-up vehicle.

△ Ensure that the axle sits securely in the saddle of the axle stand.

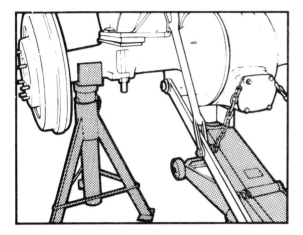

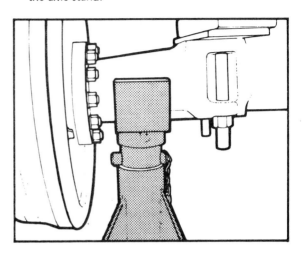

△ Take care never to exceed the maximum load of each jack and axle stand — remember that ballasted tyres, wheel weights, frame weights, etc. will add to the load to be lifted.

△ Regularly check the axle stand pins for signs of wear and replace them with new pins of correct type, if necessary.

△ Place jacks and axle stands/blocks to give support in a suitably strong position, in accordance with manufacturer's instructions.

△ Ensure that the jack stands on a solid surface.

△ Place a wide, flat plate or board beneath each jack, axle stand or block (if necessary) — to spread the load and reduce any tendency to sink into the ground.

△ Place jack well clear of any fluid lines and cables/control rods.

△ Do not use a jack operated by a tractor's hydraulic system if the tractor is fitted with frame weights, wheel weights or ballasted tyres.

△ After jacking up the vehicle, examine the jack and the vehicle for security and stability; if doubtful about its security or stability, lower the vehicle and re-position the jack before raising it again.

3. Lowering the vehicle

Ensure that all necessary tasks (e.g. locating wheel nuts) have been finished before fully lowering vehicle.

Remove all hand tools, parts, etc. from beneath vehicle before lowering it.

Raise jack slightly to remove axle stand and blocks.

Fully tighten wheel nuts after lowering.

Remove all jacks and support plates/boards from beneath vehicle and all chocks from wheels before moving vehicle away.

You should understand

— why it is important to remove any rings and wristwatch before doing any servicing work

— how to calculate the weight which is to be supported by a jack or axle stand

— why particular care is necessary when jacking up a vehicle in preparation for working on it.

If you do not understand all these points ask your instructor, manager, employer or course tutor for advice.

Further experience

You will become proficient only by practice. Try to get experience as soon as you can by taking every opportunity offered by your manager or employer to follow up the training you have received.

15 FITTING WHEEL AND FRAME WEIGHTS

Your aim

is to be able to:

- fit to a tractor or remove from it any wheel weights or frame weights required
- carry out the task in accordance with the manufacturer's recommendations
- take all precautions to minimise risks of personal injury or injury to other persons or livestock
- take all precautions to avoid causing damage to equipment or any other property.

You will need

Wheel weights or frame weights (as required).

Small tools.

Manufacturer's instruction book for the tractor.

You must

have received instruction or gained experience in:

- jacking up a tractor (trainee guide M.1.C.1)
- safe lifting and carrying (trainee guide COM.1.1).

1. Checking tyre condition

△ Examine each tyre carefully and report if damaged or in poor or unsafe condition.

Look for:

- excessive or uneven wear

- impact damage

Note △ = important safety point.

— other serious defects or damage

— adequate ply rating of the tyre for the extra loads to be added to it (if in doubt, check with tyre manufacturer or agent).

2. Fitting weights

See manufacturer's instruction book.

Fit type of weights best suited to the machine to be used and the work to be done.

If extra weight is required at the front end, use front-mounted frame weights rather than front wheel weights.

△ Lift heavy weights using safe methods (see trainee guide COM.1.1 — Safe lifting and carrying).

Always fit equal weights to left and right hand wheels.

Ensure that all weights are securely attached to tractor.

After fitting weights increase tyre pressures according to extra weight added (see manufacturer's recommendations).

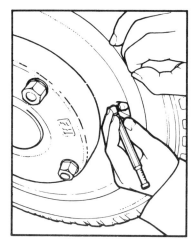

After fitting machine and weights and adjusting tyre pressures, check whether tractor is stable and will steer safely without excessive sway.

3. Removing weights

Remove weights at the same time as the machine for which they were fitted is removed from the tractor.

Do not leave weights on a tractor if there is no reason for them.

△ Lift heavy weights using safe methods (see trainee guide COM.1.1 — Safe lifting and carrying).

Check and adjust tyre pressures after removing weights (see manufacturer's recommendations).

Place all removed weights in correct storage place.

You should understand

— why it is necessary to fit extra weights to a tractor when using some machines
— why it is important to check the tyres for damage and poor condition before fitting weights to a tractor

- why it is better to use front frame weights than front wheel weights
- why extra weights should always be added in equal quantities to each side of the tractor
- why it is important to readjust the tyre pressures after fitting or removing frame or wheel weights
- relevant safety legislation.

If you do not understand all these points ask your instructor, manager, employer or course tutor for advice.

Further experience

You will become proficient only by practice. Try to get experience as soon as you can by taking every opportunity to follow up the training you have received.

16 ALTERING WHEEL TRACK WIDTH

Your aim

is to be able to:
— adjust front and rear wheel track widths
— carry out the task in accordance with the manufacturer's recommendations
— take all precautions to minimise risks of personal injury or injury to other persons or livestock
— take all precautions to avoid causing damage to any equipment

You will need

— equipment for jacking up the tractor or machine
— manufacturer's instruction book
— clean container for wheel nuts
— any tools necessary
— the help of an assistant to handle heavy rear wheels

You must

have received instruction or gained experience in:
— jacking up tractors and wheeled machines (trainee guide M.1.C.1)
— safe lifting and carrying (trainee guide COM.1.1).

Note △ = *important safety point.*

1. Jacking up tractor

Slacken wheel nuts before jacking up.

△ Jack up tractor in accordance with all the points set down in trainee guide M.1.C.1.

Jack up one end of tractor in one lift — front or rear.

2. Removing the wheels

When removing a rear wheel from the studs, adjust the height of the jack so that the weight of the wheel is not on the threads.

△ Place an axle stand so that it almost but not quite supports the axle (if the jack were to fail, the tractor would only slip a short distance before being supported by the axle stand)

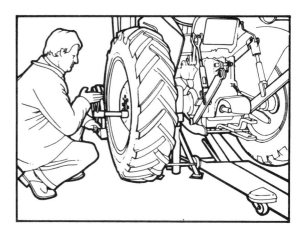

△ Remove top nuts last to prevent wheel falling on you.

Place wheel nuts in a clean container as each one is removed.

△ Always obtain assistance to handle a heavy rear wheel.

△ Always have an assistant when handling a ballasted wheel.

3. Refitting tractor wheels

Check tractor instruction book for settings to give required width.

Exchange positions of rear wheels if necessary.

Refit wheel so that tyre tread will rotate in correct forward direction.

If necessary, adjust jack height to allow wheel to be fitted on to the studs while its weight is resting on the ground.

Use correct nuts/washers.

Fit nuts with conical ends correctly on to studs.

If necessary, lift tractor so that wheel is completely clear of ground before tightening all nuts to finger tightness.

Tighten nuts in diagonal order, beginning with the top nut.

After lowering tractor, tighten all nuts to correct torque (see manufacturer's instruction book) in correct order as above.

You should understand

— why wheel track widths must be set correctly for
   — towing trailers and trailed machines
   — ploughing
   — working in row crops
   — safe working on sloping ground
   — using front-mounted machines
— why a tractor rear wheel should always be handled by two men, particularly if ballasted
— why a wheel should be fitted with the tread in the correct direction
— why all wheel nuts must be kept tightened to the correct torque.

If you do not understand all these points ask your instructor, manager, employer or course tutor for advice.

Further experience

You will become proficient only by practice. Try to get experience as soon as you can by taking every opportunity offered by your manager or employer to follow up the training you have received.

17 WATER BALLASTING TYRES

Your aim

is to be able to:

- check the condition and suitability of the tyres for ballasting
- mix antifreeze ballasting solution if necessary
- ballast the tyres
- re-inflate tyres and adjust pressures as required
- carry out the task in accordance with the manufacturer's recommendations
- take all precautions to minimise risks of personal injury or injury to other persons or livestock
- take all precautions to avoid causing damage to any equipment.

You will need

- manufacturer's instruction book for the machine
- equipment for jacking up the tractor or machine
- tyre ballasting tool fitted with the correct length tube for the size of tyre and degree of ballasting
- hose pump and water supply or a tank supported about 6' higher than the top of the tyres to be ballasted
- water containers (plastic preferably)
- tyre pressure gauge (air/water type).

You may need

- calcium chloride flakes

You should

have received instruction or gained experience in:

- all the activities in trainee guides M.1.A.1 to M.1.A.5
- jacking up tractors and wheeled machines (trainee guide M.1.C.1).
- read 'Emergency aid in case of chemical poisioning' in section 1.

Note △ = important safety point.

1. Jacking up the wheel(s)

Jack up one wheel at a time or both wheels together, according to type and quantity of equipment available for jacking.

△ Jack up wheels in accordance with all the points set down in trainee guide M.1.C.1.

2. Checking tyre condition

△ Examine each tyre carefully and report if damaged, in poor or unsafe condition or unsuitable for ballasting.

Look for:

- excessive or uneven wear

— impact damage

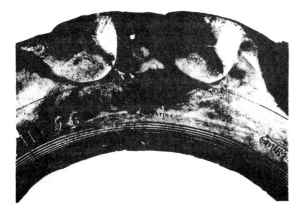

— other serious defects or damage

— adequate ply rating of the tyre for the extra load of the ballast (if in doubt, check with tyre manufacturer or agent).

3. Mixing antifreeze ballasting solution

Use plastic containers for mixing (metal containers may be corroded by the solution).

Consult manufacturer's instructions for quantity of water for each wheel and weight of calcium chloride flakes to be added to it to give the degree of frost protection required.

Do not use more water or flakes than recommended by the manufacturer.

Add the calcium chloride flakes slowly to a small quantity of the water (about 1/5 of the water), stir constantly until they have dissolved and then add the rest of the water to the solution.

△ Never add water to the flakes, always add the flakes to the water.

Allow the solution to cool before filling a tyre.

If any metal parts or objects are splashed with the solution, wash them with plenty of plain water.

Do not use common salt (sodium chloride) solution as antifreeze, unless calcium chloride is not available.

Never use calcium chloride solution in a radiator.

4. Filling the tyre

△ Only ballast a tyre when the wheel is fitted to the tractor.

Turn the wheel to bring the valve to the top and apply the parking brake.

Remove the valve core putting it in a safe clean place, allow the air to escape and fit the tyre ballasting tool.

△ Keep your eyes away from the air stream as the tyre deflates.

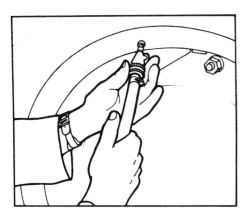

Connect the hose from the pump or gravity-flow supply tank.

Pump in the solution, or allow it to flow into the tyre from the supply tank.

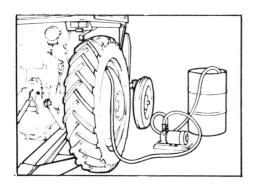

If applicable, at intervals during filling, shut off the solution and release trapped air by pressing the deflator button on the ballasting tool.

The tyre is filled to the correct level when solution comes out of the bleed hole (or deflator button when pressed).

△ Do not fill a tyre to a greater degree than the maximum recommended by manufacturer.

△ Do not completely fill a tyre.

Remove the ballasting tool, replace the valve core, rotate the wheel to place the valve at the bottom and inflate tyre to normal working pressure for a ballasted tyre.

Wash the ballasting tool, pump and hoses thoroughly in plenty of clean water.

Lower the jack and re-check the tyre pressure, adjusting to required pressure if necessary.

5. Checking the pressure of a ballasted tyre

See manufacturer's instructions for correct tyre pressure.

Use an air/water pressure gauge to check a ballasted tyre.

Check pressure of a ballasted tyre only when the valve is at the bottom of the wheel.

Check pressure of ballasted tyres more often than non-ballasted tyres (at least 2-3 times each week).

△ Do not over-inflate a ballasted tyre.

Do not reduce tyre pressure below the recommended minimum.

6. Emptying the ballasting solution from a tyre

△ Ballast should be emptied from a tyre before the wheel is removed. If not, always have at least two men to handle a wheel with a ballasted tyre.

Jack up wheel (see trainee guide M.1.C.1.) using chocks and axle stand.

Turn wheel to bring the valve to the top and apply the parking brake.

Release the air pressure and remove the valve core, putting it in a safe, clean place.

Release the parking brake and rotate the wheel so that the valve is at the bottom.

If required for re-use, collect the solution in clean plastic containers as it runs out. (A tyre of 18" section may contain as much as 85 gallons of solution).

When the solution no longer runs out, refit the valve core and re-inflate the tyre to the required pressure.

Do not allow the liquid to contact any metal parts — although not dangerous it may cause serious corrosion.

Any splashes of solution should be washed off with clean water, since any surface coated with calcium chloride will remain damp.

At all times keep the solution away from the electrical system.

(Note that some solution will always remain in the tyre unless the tube is removed from the tyre).

If the solution is to be kept for future use, label the container clearly before storing away.

Dispose of any solution no longer required, washing it away with plenty of clean water.

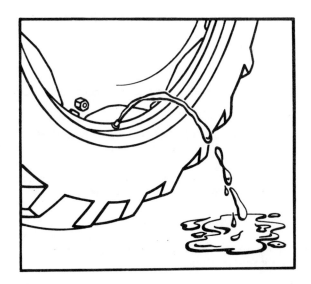

You should understand

- why it is important to check the tyre for poor condition and damage before it is ballasted

- why it is dangerous to add water to flakes of calcium chloride

- why any metal parts which become wetted with calcium chloride solution must be thoroughly washed with clean water

- why a tyre must never be completely filled with ballasting solution

- why the valve must always be at the bottom of the wheel when checking the pressure

- why a tyre must never be over or under-inflated

- why you should empty the ballast from a tyre before removing the wheel from the tractor, or why at least two people should handle a wheel which must be removed without being emptied.

If you do not understand all these points ask your instructor, manager or employer.

Further experience

You will become proficient only by practice. Try to get experience as soon as you can by taking every opportunity to follow up the training you have received.

SECTION 3: How to maintain your tractor

The good operator looks after his tractor and will often prefer to maintain it himself, rather than let someone else do it for him. He is aware that 'prevention is better than cure', and knows that preventive maintenance will prolong the life of his tractor, and save money by reducing breakdowns and expensive repairs.

Because every different model of tractor has slightly different maintenance requirements, the guides in this section rely heavily on the manufacturer's operator's manual. The manual should be regarded as essential for all but basic maintenance operations.

The most important aspects of engine maintenance which is not covered in this section is servicing the diesel fuel system. Apart from the necessity of filling the fuel tank after work with clean fuel, and removing any water which has accumulated in the sediment bowl, the operator should not attempt any maintenance on the system without instruction from a competant person. A very tiny amount of dirt entering either the fuel pump or the injectors can do a lot of damage, requiring expensive repairs so this aspects of maintenance is best left to an expert until appropriate training has been undertaken by the operator.

18 SERVICING TRACTORS (EXCLUDING ENGINES)

Your aim
is to be able to:
- grease and oil the tractor
- adjust the brakes
- road test the brakes
- check the clutch pedal clearance and adjust if necessary
- top up the steering box and/or the power steering reservoir
- check free play in front wheel hub bearings
- inspect tyres and check pressures
- top up and change hydraulic oil and oil in gear box and final drive
- road test the steering
- carry out all the tasks in accordance with the manufacturer's recommendations
- take all precautions to minimise risks of personal injury or injury to livestock
- take all precautions to avoid causing damage to equipment or any other property.

You will need
- equipment for jacking up the tractor
- manufacturer's instruction book for the machine
- any tools necessary
- a firm level site
- a flat surfaced area, sufficiently large to carry out braking and steering tests
- sufficient oil of correct types/grades for topping or changing oil in steering box, gearbox and final drive, and hydraulic oil.
- equipment for measuring toe-in of front wheels
- replacement tyre valve cores and valve core extractor
- airline incorporating pressure gauge
- grease and oil guns containing the recommended grades/types of grease and oil
- cleaning rags (not cotton waste).

You must
Have received instruction or gained experience in:
- driving around the holding (trainee guide M1.A.4)
- jacking up tractors and wheeled machines (trainee guide M1.C.1).

Note: △ = *important safety point*.

1. General safety

△ Position a tractor on level ground for servicing.

△ Stop tractor engine before doing any servicing operations and remove key.

△ Remove wristwatch and any rings or bracelets before doing any servicing operations.

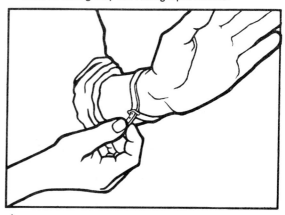

△ If it is necessary to work on a tractor which has a fore-end loader fitted, lower the loader to the ground if possible or, if beam must be kept in raised position, on a firm support (e.g. a wall).

2. Topping up and changing oil in transmission

Use the correct type/grade of oil (see manufacturer's instructions).

Follow manufacturer's recommendations on frequency of oil changes.

Clean or change filters, as recommended, replacing sealing rings.

Allow oil to drain out for at least five minutes.

Clean drain plug and replace when ready to refill.

Examine old oil for foreign bodies and water.

Clean around filler plugs and caps before removing and refilling.

Dispose of old oil safely.

Fill or top up to correct level:
— with new oil
— using clean container for handling oil
— without over-filling.

3. Greasing or oiling all points not covered in daily checks

Use the correct type/grade of grease or oil (see manufacturer's instructions).

Lubricate each point as often as is recommended by manufacturer.

Clean grease gun nozzle, if necessary, before applying to each point.

Clean each grease nipple before and after lubrication.

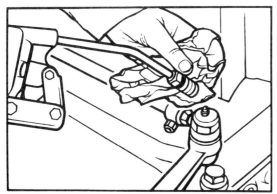

Inject the correct amount of grease into each point (see manufacturer's instructions).

Always lubricate the points in the same order.

If a grease point will not take the grease, find out the cause and remedy the defect before continuing.

Replace any damaged grease nipples with new ones.

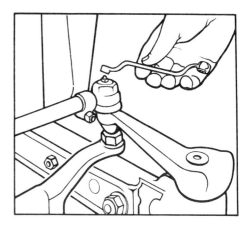

4. Jacking up the vehicle

Follow the points set down in trainee guide M1.C.1, 'Jacking up tractors and wheeled machines'.

5. Servicing the clutch

Check the clutch pedal free travel and adjust or report condition if necessary.

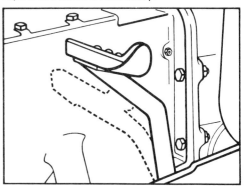

Adjust the clutch in accordance with the manufacturer's instructions.

6. Servicing the steering

(a) Top up steering box and/or power steering reservoir, if necessary, in accordance with manufacturer's instructions
— use the correct grade/type of oil.

(b) Check front wheel hub bearings for free play and report condition or adjust if necessary. (See trainee guide M4.E.1 for details of adjustment).

(c) Measure, and adjust if necessary toe-in of front wheels. Measure toe-in of front wheels between similar positions on inside rims.

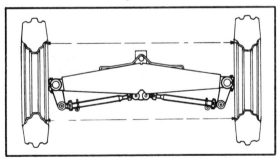

IH – TA514 – 40 B

After taking first measurement, rotate each wheel a half turn and take second measurement (or as recommended by manufacturer).

Adjust toe-in in accordance with manufacturer's instructions.

7. Checking tyres

Visually inspect all 4 tyres looking for:

— cuts and bulges

— cracks

— any point where canvas shows

— state of wear, and wear characteristic.

You will need to move the tractor to inspect all the way round the tyre.

Discuss tyre condition and wear with your supervisor/manager.

Check tyre pressures and reset to correct pressure.

Check tyre valve cores for leaks by moistening with saliva and watching for bubbles.

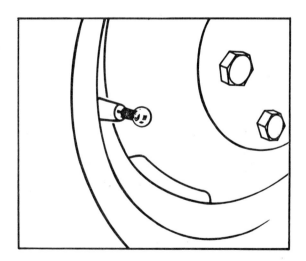

— replace any core which leaks.

8. Adjusting the brakes

△ Adjust the brakes in accordance with the manufacturer's instructions.

△ After adjusting brakes, ensure that independent brake pedals have equal travel.

△ Always road test brakes after they have been adjusted.

9. Testing the brakes

△ Test the brakes for efficiency and balance and report condition if necessary.

△ Road test the brakes off the public highway.

△ Test the brakes on a flat surface.

△ Carry out the initial tests at low speed.

△ Only test the brakes when the front wheels are pointing straight ahead.

10. Testing the steering

△ Road test the steering off the public highway.

△ Test the steering on level ground at low speed.

Recognise wear in steering by excessive movements of the steering wheel necessary to maintain a straight course.

You should understand

— why it is important to remove any rings and wristwatch before doing any servicing work

— why each grease nipple must be cleaned before and after injecting the grease

— why independent foot brakes should be adjusted so that the pedals have equal travel

— why brakes should only be tested if the vehicle is travelling slowly and in a straight line

— why a used split pin should not be refitted to the castellated nut on a front wheel hub bearing

— why front wheel toe-in must be accurately set

— why it is important to inspect tyres regularly and carry out repairs immediately

— why it is important to keep detailed records of servicing work which you have carried out

— relevant safety legislation.

If you do not understand all these points ask your instructor, manager, employer or course tutor for advice.

Further experience

You will become proficient only by practice. Try to get experience as soon as you can by taking every opportunity offered by your employer or manager to follow up the training you have received.

19 ENGINE LUBRICATION

Your aim

Is to be able to:

- change the engine oil
- clean or change the oil filter
- carry out other lubrication tasks, all in accordande with the manufacturer's recommendations
- take all precautions to minimise risks of personal injury or injury to other persons or livestock
- take all precautions to avoid causing damage to any equipment.

You will need

Manufacturer's instruction book.

Lubricating oil of correct types/grades for the engine, gearbox, final drive and fuel injection pump.

Clean utensils for handling oil.

Container for used oil.

Replacement elements for oil filter and hydraulic filter.

Any tools necessary.

Cleaning rags (not cotton waste).

You must

have received instruction or gained experience in:

- manoeuvring the machine around the holding. where appropriate.

Note △ = important safety point.

1. General safety

△ Position machine so that engine is level before servicing.

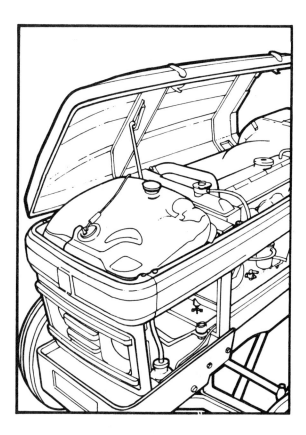

△ Stop engine before doing any servicing operations (except when checking lubricating and cooling systems after refilling, or when road-testing).

△ Remove wristwatch and any rings or bracelets before doing any servicing operations.

△ If it is necessary to work on a tractor which has a fore-end loader fitted, lower the loader to the ground if possible, or, if beam must be kept in raised position, on a firm support (e.g. a wall).

2. Changing/checking engine oil etc.

Change/check level of oil by the recommended method, as often as recommended by the manufacturer.

Warm up the engine before draining the oil.

Leave drain plugs out until almost ready for refilling with oil, or allow to drain for at least five minutes.

Examine old oil for excess water and any foreign bodies and report if necessary.

After draining oil, do not wash out sump with paraffin or similar cleaning liquid

Clean crankcase breathers, if required.

Clean drain plugs before replacing.

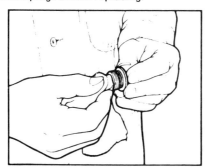

When changing oil filter, replace the sealing ring also.

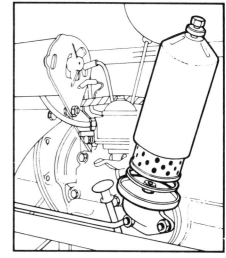

Clean around a filler cap before removing it and refilling.

Always use clean, new oil for refilling.

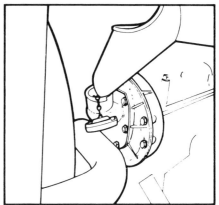

Refill with correct type/grade of oil (see manufacturer's instructions).

After refilling, replace filler cap, start engine, check for correct oil pressure and check to see if any oil leaks can be seen. If a leak is found check cause, stop engine and repair, or report to supervisor/manager.

Dispose of or use the old oil in a suitable manner.

You should understand

- why you should remove rings and wristwatch before doing any servicing work
- why it is important to keep lubricating oil as free from dirt as possible
- why it is better to warm up an engine before draining the oil
- why it is important to keep detailed records of servicing work which you have carried out.
- why a fore-end loader should be lowered to the ground, or firmly supported, before you work on the tractor.

If you do not understand all these points, ask your instructor, manager, employer or course tutor for advice.

Further experience

You will become efficient only by practice. Try to get experience as soon as you can by taking every opportunity offered by your manager or employer to follow up the training you have received.

20 SERVICING COOLING SYSTEMS

Your aim

is to be able to:
- drain and flush out the cooling system
- mix a fresh coolant solution using anti-freeze or corrosion inhibitor as required
- refill the cooling system correctly
- check, adjust and replace the fan belt and hoses
- clean out the radiator core
- carry out all the tasks in accordance with the manufacturer's recommendations
- take all precautions to minimise risks of personal injury or injury to livestock
- take all precautions to avoid causing damage to equipment.

You will need

- manufacturer's instruction book for the machine
- sufficient anti-freeze/corrosion inhibitor to prepare coolant for complete system
- graduated measure
- replacement fan belt of correct size/type for the engine
- radiator flushing compound
- warning notices (system drained, anti-freeze, corrosion inhibitor)
- any tools necessary
- cleaning rags (not cotton waste).

You must

have received instruction or gained experience in:
- driving round the holding (trainee guide M.1.A.4).

Note △ = important safety point

1. General safety

△ Follow at all times the safety points set down in section 1 of trainee guide on Daily Checks (M.1.A.1).

2. Draining and flushing the cooling system

Check manufacturer's special instructions before draining a sealed cooling system.

△ Do not remove a radiator cap from a hot engine.

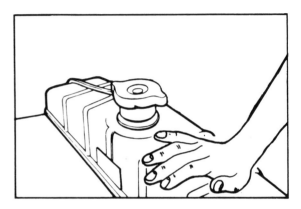

Open all drain taps.

Ensure that radiator overflow pipe is not blocked.

Use a recommended flushing compound and/or back flush with mains water pressure

Check all water hoses for condition and replace any showing signs of deterioration.

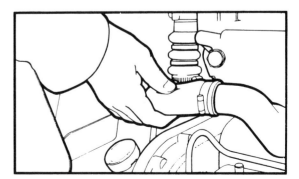

Check radiator cap seal for good condition and spring for correct operation.

— replace with a new cap of correct type, if faulty.

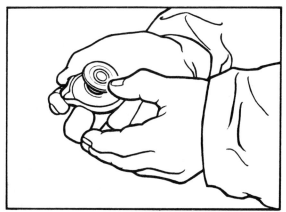

3. Preparing coolant and refilling system

Use anti-freeze or corrosion inhibitor, as required, of a type recommended by the engine manufacturer.

Use clean soft water when preparing the mixture.

Mix the coolant according to the manufacturer's instructions to produce the correct strength for the degree of protection required.

Fit warning notices to a drained system or a system containing anti-freeze or corrosion inhibitor.

Do not use tyre ballasting solution (i.e. water containing calcium chloride) in a cooling system.

Do not overfill header tank (see engine manufacturer's instructions)

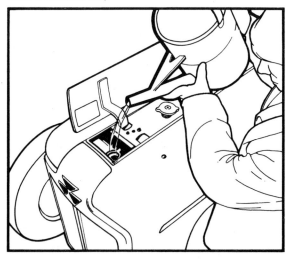

After refilling header tank to correct level, run engine for five minutes and then top up with more solution, if necessary, to ½" below filler cap (or as stated in manufacturer's instructions)

After refilling replace radiator cap and, with the engine running, check all the joints in the system for leaks, and tighten clips where required.

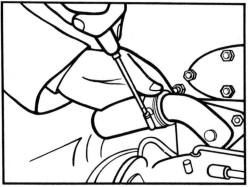

It is preferable to top up a cooling system, when this is necessary, with a pre-mixed coolant solution of the correct strength.

4. Checking and adjusting fan belt and cleaning radiator core

△ Remove switch key before checking, adjusting or removing fan belt.

Check fan belt condition and replace if necessary with a new belt of the correct type for the engine.

△ Always keep your fingers on the outside of the belt when replacing it.

Set the fan belt to the correct tension (see engine manufacturer's instructions.

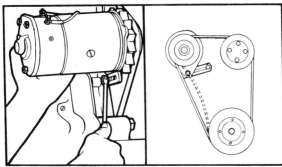

△ Do not test or reset the tension with the engine running.

△ Only clean the radiator core when the engine is stopped.

Clean the radiator core from the engine side of the radiator.

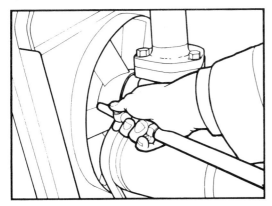

You should understand

— why you should remove any rings and wristwatch before doing any servicing work

△ — why it is important to take special precautions when removing the radiator cap from a warm or hot engine

— why a cooling system containing anti-freeze or corrosion inhibitor should not be topped up with plain water

— why it is important to keep detailed records of servicing work which you have carried out.

If you do not understand all these points ask your instructor, manager, employer or course tutor for advice.

Further experience

You will become proficient only by practice. Try to get experience as soon as possible by taking every opportunity offered by your employer or manager to follow up the training you have received.

21 SERVICING VEHICLE ELECTRICAL SYSTEMS

Your aim

is to be able to:

- remove the battery, charge it and refit it
- check the electrical fittings
- identify and replace blown fuses and spent light bulbs
- carry out all tasks in accordance with the manufacturer's recommendations
- take all precautions to minimise risks of personal injury or injury to other persons or livestock
- take all precautions to avoid causing damage to any equipment.

You will need

- manufacturer's instruction book for the machine
- battery charger
- distilled water
- petroleum jelly
- any tools necessary
- cleaning rags (not cotton waste)
- first aid equipment which complies with the Health and Safety (First Aid) Regulations, 1981.

You may need

- supply of boiling water.

You should

have received instruction or gained experience in:

- all the activities in Unit A of Wheeled tractor operation
- servicing cooling systems (trainee guide M.4.A.2).

Note △ = *important safety point*

1. General safety

△ Follow at all times the safety points set down in section 1 of trainee guide M1.D.1 (servicing tractors — excluding engine) or section 1 of trainee guide M.4.A.1 (engine lubrication).

△ Never place tools etc on top of a battery.

2. Servicing the battery

△ Handle a battery with care at all times and do not knock or bump it.

△ Keep battery level at all times.

△ Do not smoke or use a naked light when inspecting level of fluid in the cells of a battery.

△ If battery fluid is spilled on hands or clothing, wash it off at once.

Use a charger according to manufacturer's instructions.

△ Always disconnect charger from mains before connecting or disconnecting a battery.

Remove battery cell covers (or screw caps) when charging, unless instructions on battery state otherwise.

Set charge rate of charger to level recommended by battery manufacturer.

△ Do not smoke or use a naked light near a battery which is being charged.

Clean powdery deposits from a battery terminal with boiling water.

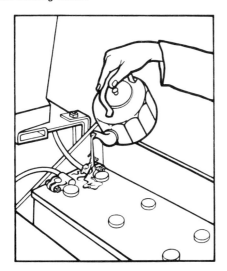

Before refitting battery connections to battery, grease terminal posts and clamps with a layer of petroleum jelly to prevent corrosion. Ensure that all metal surfaces of the terminals are well covered after refitting.

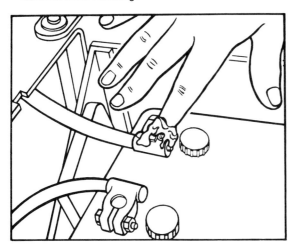

△ Avoid arcing across battery terminals with metal tools.

3. Servicing electrical fittings

Disconnect battery earth cable before working on any part of the electrical system.

Look carefully for loose or dirty terminals and fraying, burnt or broken wires.

Check horn, lights, wiper etc. for correct operation. Adjust if necessary and replace spent bulbs and any cracked or broken lenses (see manufacturer's instruction book for details).

Lubricate generator bearing as recommended by manufacturer (if applicable).

△ Avoid arcing across battery terminals with metal tools.

Ensure that the battery leads are always replaced on the correct battery terminals, i.e. with similar symbols placed together. (Note that P is the same as Positive, or +) (This is particularly important if the engine has an alternator, which will be seriously damaged by wrong connection.)

If the battery has been removed or disconnected, do not attempt to tow start a machine which has an engine with an alternator.

Never use an electric welder on an implement which is in contact with any part of a tractor fitted with an alternator without first disconnecting the alternator from the electrical system.

4. Servicing the dynamo/alternator

Check fan belt tension and adjust if necessary (see trainee guide M.4.A.2.— servicing cooling systems).

Lubricate dynamo/alternator as required (see manufacturer's instruction book for details).

You should understand

— why you should remove any rings and wrist-watch before doing any servicing work

— why it is dangerous to smoke or use a naked light near a battery

— why it is important to keep detailed records of servicing work which you have carried out.

If you do not understand all these points ask your instructor, manager or employer.

Further experience

You will become proficient only by practice. Try to get experience as soon as you can by taking every opportunity offered by your employer or manager to follow up the training you have received.

SECTION 4: How to carry out field operations

The final section deals with some of the field operations which a tractor operator will be asked to carry out from time to time. Some of them may not be relevant to the reader of this book at present: if, for example, there is not a fore-end loader on his tractor there is no immediate need to study the first guide. If, however, there is a disc plough on the farm, it can still be useful to study the guides on mouldboard ploughing — the method of use is similar even if the adjustments are different.

The guides on seed drill maintenance can be applied to most drills, and those on mowers to several types of harvesting machinery.

The two guides on crops sprayers are extremely important because of the need to apply exact quantities of chemical to the crop. Not only is it expensive to waste chemical by applying the wrong amount, but it may also be ineffective and harm the crop or the environment.

22 OPERATING A FORE-END LOADER

Your aim

is to be able to attach and detach a front end loader and operate it efficiently to move and load material:

- observing all safety regulations and precautions to minimise risk of injury to persons or damage to buildings and equipment
- keeping it in good operating condition
- with minimum wear and tear and breakdowns.

You will need

- a level area of concrete on which to attach the loader
- manufacturers' instruction books for both tractor and loader
- tractor fitted with radiator guard, loader sub-frame and ram units
- fore end loader beams and bucket or fork attachments
- tools as required
- rear weight blocks
- trailer
- material to be loaded
- first aid equipment which complies with the Agriculture (First Aid) Regulations 1957.

You must have

received instruction or gained experience in:

- manoeuvring with a two wheeled trailer or trailed machine (M.1.B.6)
- obtaining maximum traction
- altering widths of wheel track (M.1.C.4)
- water ballasting tyres (M.1.C.5)
- fitting wheel and frame weights (M.1.C.2).

Note: △ = important safety point.

1. General safety

△ Always stop engine, apply parking brake a put all controls in a neutral position with bucket lowered, before getting off tractor.

△ Never stand under fork or bucket or allow anyone else to do so.

△ Fasten any loose or flapping clothing before operating or working on loader.

2. Introduction

Remember when operating the loader that when a load is carried at the front, weight will be taken off the rear wheels which will affect traction and the stability of the tractor.

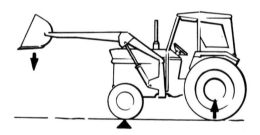

3. Fitting loader beams

Remove any front end or front wheel weights.

Adjust wheel track as necessary.

Check tyre pressures and adjust according to manufacturer's recommendations.

Align tractor with beams:

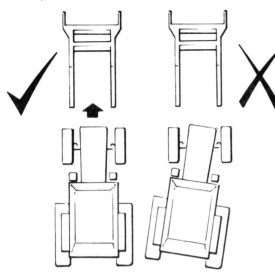

- carefully approach beams at a speed which allows you to stop quickly and easily
- lower rams from stowage position; avoid dropping or damage may result
- manoeuvre carefully between beams to align holes in beams and sub-frame
△ — do not use fingers to align holes, use suitable bar, to avoid injury if the beams should drop
- insert and secure all locating and linch pins
- connect trip, either cable or hydraulic
- stow jack stands or, if removed, store in correct place

4. Attaching bucket fork

When attaching bucket/fork ensure that:
- bucket is of the correct size for material to be handled if a choice is available
- locating pins or latching device are secure.

When fork is attached:
- check tightness and straightness of fork tines tightening and straightening as necessary
- check operation of trip mechanism.

Connect return ram or spring, ensuring that:
- all connections are secure
- hydraulic hoses are undamaged and correctly routed
- external hydraulic control valve is in live position.

5. Attaching rear weight block (where necessary)

Attach weight to link arms ensuring that:
- drawbar is stowed
- top link is removed and stored
- correct bar is used for weight.

Align block with bar:
- ensure block is correct for work to be done
△ — do not attempt to manhandle block to obtain correct alignment.

- reversing carefully, manoeuvre into position.

Lift block on hydraulics and check that:
- linkage hydraulics are isolated
- rear wheels are not fouled (adjusting check chains if necessary)
△ — feet are not underneath block when adjusting check chains
- block is securely attached
- locking pins are secure.

6. Checking operations

Ensure that all moving parts are lubricated.

Get on tractor, start engine and check:
- operation of all controls
- each ram to full extent of movement

7. Maintaining stability

Causes of lateral instability
— exceeding maximum safe working load
— turning at speed
— operating with a soft tyre

— uneven ground.

Causes of longitudinal instability

— overloading
— rear weight block too light
— violent braking
— jerky operation of hydraulics, particularly when using high lift.

Carry loads as low as possible.

Drive at the correct speed for conditions:
— slow down for slippery, loose or uneven surfaces.

Operate controls smoothly, avoiding:
— fast starts
— jerky stops
— sudden turns.

Never exceed the recommended capacity of the loader which will be reduced by:
— extending the reach of loader

— adding attachments.

When travelling up or down slopes keep load uphill at all times.

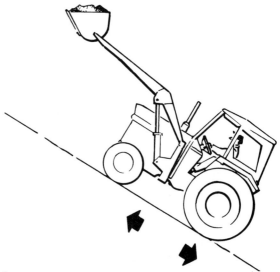

△ Never turn or drive across a slope with bucket in the raised position.

Always drive across furrows at right angles to line of furrows.

8. Safe driving

When driving tractor with loader attached, always:
- latch independent brakes together
- take extra care when operating near overhead cables or trees, or when passing through doorways
- allow for bucket swinging out particularly when reversing
- ensure that your forward view is not obscured by the bucket

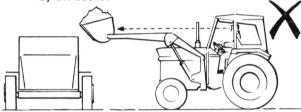

9. Filling trailer

(a) *When loading:*
- ensure that trailer is on hard level ground, where possible with a firm base for the drawbar jack or shoe
- park trailer to ensure the minimum of turning and travelling for the loader

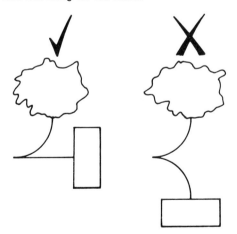

- park trailer securely; place blocks behind and in front of wheels

- apply trailer brake
- whenever possible carry out loading on a level area
- do not approach trailer across the line of a slope
- maintain stability by keeping the load low on sloping ground.

(b) *When filling bucket/fork:*
- set hydraulic quadrant marker in position at which loader is neither lifting or lowering
- select suitable gear and engine speed; this will depend upon:
  (i) type of surface; rough or smooth, hard or soft
  (ii) amount of room available for manoeuvring
  (iii) material being loaded
- approach stack at a speed just sufficient for bucket/fork to be driven into face of material by tractor's forward motion
- remove material from top of heap first and work downwards
- ensure that bucket/fork is lifting as it enters face of material
- increase engine revs, using foot throttle if fitted, to increase speed of lift
- reverse away from heap.

(c) *When emptying bucket or fork ensure that you know the maximum load for the trailer.*

Position bucket/fork loads, to fill:
- trailer from centre with an evenly distributed load
- rotary spreaders from both ends
- moving floor spreaders from end opposite to beaters.

Avoid:
- lifting load higher than necessary
- tipping heavy loads of material from excessive heights

- riding clutch when manoeuvring
- approaching trailer at a speed too high for complete control
- fouling loader, bucket, fork or tines on any part of trailer
- overloading trailer, particularly with heavy materials

Lower bucket/fork to ground when:

- servicing tractor
- leaving tractor unattended.

△ If beams must be raised for servicing ensure they are firmly supported.

10. Removal and storage of loader

When work has finished for the season:
- clean loader particularly around trip mechanism
- lubricate all moving parts
- apply waste oil or anti-corrosive agent to all bright parts.

(a) *To remove rear weight block:*
- re-activate linkage hydraulics
- lower block to ground where it is to be stored
- detach block and secure locking pins
- do not attempt to manhandle block.

(b) *To detach bucket/fork:*
- lower bucket/fork to ground in its storage position
- release latching device/locating pins
- stow locating pins in holes in bucket
- disconnect return spring/ram (as applicable)
- lift loader slightly, if necessary, to release trip
- reverse slowly and carefully until well clear of bucket/fork.

(c) *To detach loader beams from tractor:*

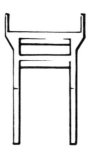

- ensure that stands are secure and will support detached beams
- ensure that trip cable or hydraulic trip lines are disconnected
- set external hydraulic control valve in neutral position (where applicable)
- detach rams from beams, ensuring they do not drop onto front axle
- stow locating pins in beams
- detach beams from sub-frame uprights and stow locating pins
- reverse slowly and carefully away until well clear of beams

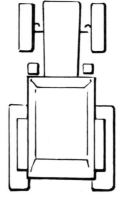

- secure rams on sub-frame
- check loader is securely parked and all components are stored together in correct place.

11. Driving on the highway

When operating on the highway:

- ensure wheels are clean

△ — position bucket as low as possible above eye level without obstructing vision

- ensure bucket is neither rising nor falling

△ — do not allow passengers on tractor or in bucket

△ — ensure that loader does not project into stream of cross traffic at road junctions or gateways onto roads

△ — take special care that any load is secure when driving at road speeds, turning and braking.

You should understand

- why immunisation against tetanus is advisable
- how the stability of the loader is affected by the way in which it is driven and its load is carried
- why positioning of trailer relative to heap is important
- abrupt stops, starts and manoeuvring will cause unnecessary wear and tear to tractor.

Further experience

You will become proficient only by practice. Try to get experience as soon as you can by taking every opportunity offered by your manager or employer to follow up the training you have received.

Remember that the Board and your county agricultural college offer a wide range of courses in practical skills, together with associated knowledge, which will help you understand more about the work you are doing.

23 PLOUGHING WITH A MOUNTED RIGHT-HAND MOULDBOARD PLOUGH

Tools and materials check list

Tractor (suitable for plough being used)
Tractor instruction book
Plough
Plough instruction book and parts list
Tool kit
Clean rags
Grease gun
Oil can
Steel tape measure
Shares and bolts
Marking out poles
2m straight edge
Tyre pressure gauge

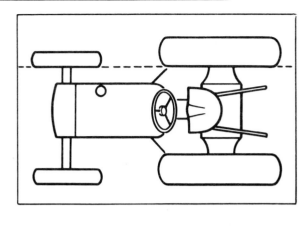

(ii) Inside of righthand front and rear wheels in line (more wear on front tyre wall with this method).

(iii) An intermediate position.

Tyre pressure

Refer to tractor instruction book usually —
front — 2.0 bar
rear — 0.8 bar

Preparation of tractor

Wheel track widths

See plough instruction book for recommended rear wheel track centres. Three methods of setting front wheels.

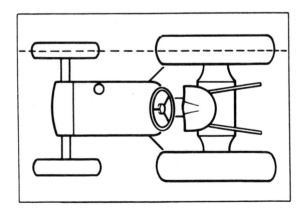

(i) Centres of righthand front and rear wheels in line;

Linkage

Correct category to suit plough. Refer to tractor and plough manufacturers' instruction books for correct linkage arrangement for category and implement concerned, and correct setting of stabilisers and check chains.

Lift rods

Set lefthand rod to correct length as recommended in the plough instruction book. Ensure that enough thread is engaged on both rods to lift plough.

Hydraulic system

Normally draught control for mounted ploughs without depth wheels.

Preparation of plough

Alignment

Landside
Remove the plough shares and use a straight edge against the landsides. The bodies should be parallel to within 5mm in 1m. A and B should equal the furrow width of the plough as set.

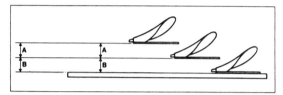

Share
After fitting new shares, the tips should all touch the ground and be in a straight line, and equidistant.

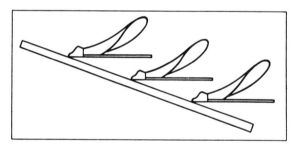

Mouldboard
Mark at a fixed distance from the front. Measure the distance between mouldboards at these points. If the spacings A are not identical, some adjustment may be made by adjusting the stays to make them as nearly as possible identical. The distance at A should equal the distance at B.

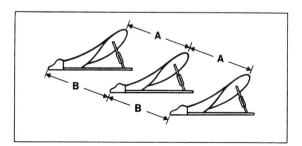

Mounting

Both crank(s) of cross-shaft should be vertical. Adjust screw lever so that collar is half way along the thread.

Order of attachment of links: Lower left, Lower right, Upper.

Body Pitch

This is checked by measuring distance from share point to plough beam with new shares on. Should be the same for each body, and as recommended in instruction book (see instruction book for method of adjustment).

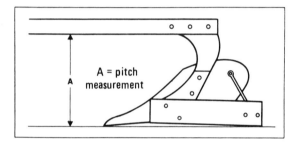

Lubrication

Main points:　Crosshaft bearings
　　　　　　　Crosshaft lever thread
　　　　　　　Disc coulter bearings.

Marking out

Measuring headland width

Allow twice length of tractor and plough, or enough space to turn tractor and plough easily. Better too wide than too narrow.

Marking out headlands

Either

follow stubble lines or combine wheelings where possible

Or

use assistant and length of rope.

Upper link extended.

Use rear furrow.

Cutting edge of disc coulter level with underside of share.

Turn furrow towards land (can turn away from land, but less tidy finish).

Lift at Corners.

Set in again at the required angle.

Marking out ridges

Allow about 10m per furrow for land width.

Plough the field along its length where possible.

Use marking out poles.

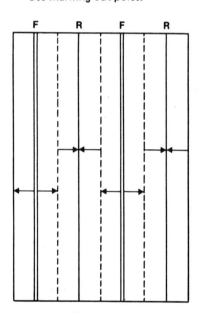

R = Ridge
F = Furrows

Opening up

Arable

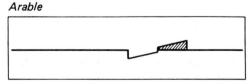

First run — rear body cuts a shallow furrow; rear disc level with bottom of share.

Second run — slightly deeper; furrow turned away from the first.

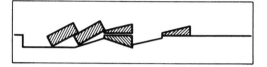

Third run — rear coulter in normal position.

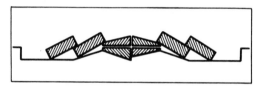

Fourth run — the ridge is finished. Front furrow slightly deeper.

Grassland

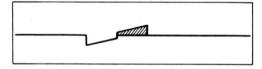

First run — rear body cuts a shallow furrow; rear disc level with bottom of share.

Second run — front body shallow.

Third run — the ridge is finished.

General Note

To avoid frequent alterations of settings, complete all first runs, then all second runs, then all third runs.

The arable opening can also be used on grassland in certain circumstances.

Ploughing in lands

Basic adjustments should be done in this order:

Depth

Adjust lever on tractor hydraulic system (or depth wheel where used).

Level

Adjust with levelling lift rod. If viewed from the rear, plough frame should be parallel with ground.

Pitch

Adjust upper link.

Landside of rear body should lightly mark the furrow bottom.

If viewed from the side in motion, plough frame should be parallel with ground.

Front furrow width

Adjust crosshaft, width should be the same as other furrows.

Disc coulters

Preliminary settings should be:

(a) Hub of disc over point of share.

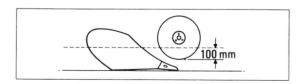

(b) Disc cutting up to 100mm into the land, depending on conditions.
(c) Hub of disc not fouling ground.

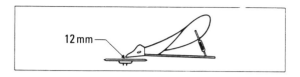

(d) Cutting edge of disc coulter 12mm clear of side of share when disc is parallel to landside.
(e) Angle normally vertical but:

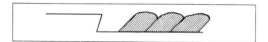

undercut assists burying trash — remove skim coulters.

overcut gives furrow wall on loose land — remove skim coulters.

Skim coulters . . .
. . . should be set:

(i) Just deep enough to cut a continuous skim and bury all trash.
(ii) With the tip as close as possible to the disc without actually touching it.

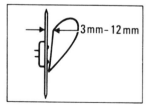

Skim coulter viewed from rear

(iii) With the heel from 3mm to 12mm clear of the disc, depending on type of plough and conditions.
(iv) With the tip below the disc rivets, and skimming off the required amount.

Finishing

Alter front furrow widths as necessary to make furrow walls parallel for the finish. Before starting the finish, the land should measure twice the full cut of the plough less one furrow width.

e.g. for three furrow plough

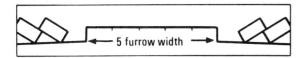

Before starting finish, start shallowing up.

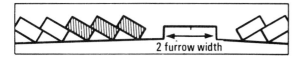

Continue shallowing.

Rear body should just skim a shallow crumb furrow.

Ploughing out headlands

Plough away from the hedge one year and towards it the next, where possible.

FAULT ANALYSIS

Fault	Cause	Remedy
1. Poor penetration	Ground too hard and dry for ploughing.	Ideally wait until ground conditions improve. The remedies below may help where it is essential to plough now.
	Worn shares.	Replace worn shares with a new set. Increase bar-point extension, or reverse or replace worn bar-point shares. Save the worn shares for easier conditions. Steel shares may be rebuilt by a blacksmith.
	Insufficient pitch on plough bodies.	Where individual adjustment exists — increase pitch on each body by the same amount.
	Disc coulters causing plough to ride out.	Raise disc coulters slightly. Move disc coulters back so that share penetrates before disc cuts.

Fault	Cause	Remedy
1. Poor penetration — *continued*	Plough tilted backwards 'taildown'.	Mounted ploughs: Shorten upper link. Trailed ploughs: Raise hitch point on plough.
	Lower links 'bottoming' — lift rods short.	Lengthen lift rods.
2. Plough will not run shallow	Ground too soft and wet for ploughing.	Wait until conditions improve.
	Too much pitch on plough bodies.	Decrease pitch on individual bodies by the same amount — where individual adjustment exists.
	Plough tilted forward — 'tail high'.	Mounted ploughs:— Lengthen upper link. Trailed ploughs:— Lower hitch point on plough.
	Bar-point shares extended too much.	Reduce bar-point extension.
	Depth wheel follows tractor wheel in soft land.	Adjust position or setting accordingly.
	Depth wheel sinking in ground.	Fit depth wheel with broader rim. Plough when conditions are more suitable or use draught control.
	Disc coulters set too far back and not deep enough.	Move coulters forward and deeper to support more of the plough weight.
	Too much weight carried on the plough.	Remove.
	Draught control insufficiently sensitive.	Fit upper and/or lower links in alternative positions to enable increased sensitivity. Reset draught control system accordingly.
3. Variations in ploughing depth	Hard and soft patches in the land.	Set flow control of draught control hydraulics slow.

Fault	Cause	Remedy
3. Variations in ploughing depth – *continued*	Draught control insufficiently sensitive.	Fit upper and/or lower links in alternative positions to enable increased sensitivity. Reset draught control system accordingly.
	Too much pitch on plough bodies.	Decrease pitch on individual bodies by the same amount – where adjustment is available.
	Plough tilted forward – 'tail high'.	Mounted plough – Lengthen upper link. Trailed plough – Lower hitch point on plough.
	Plough frame distorted.	Check as in 'Plough Overhaul.'
4. Ploughing on adjacent bouts fails to match up	Front furrow too wide or too narrow.	Avoid steering inaccuracies – drive with tractor rear wheel close to furrow wall. Make minor adjustment – (up to 50 mm) by rotating the cross-shaft. Make larger adjustments by sliding the cross-shaft across *or* altering the tractor track widths.
	Front furrow too deep or too shallow.	Check whether the legs to the plough bodies are perpendicular to the ground – check from the rear of the plough. *If they are not:* Shorten or lengthen right-hand lift rod via levelling box until they are. *When they are:* Lengthen or shorten the upper link until correct front furrow depth is achieved.
	Front mouldboard out of adjustment. Rear mouldboard out of adjustment.	Check that front mouldboard and others are adjusted equally.
	Front or rear disc coulter set too wide or too narrow.	Check settings and reset as necessary. All disc coulters should be set equally.

Fault	Cause	Remedy
4. Ploughing on adjacent bouts fails to match up – *continued*	Tractor furrow wheel sinking, causing front furrow to appear smaller or lower than the rest.	Increase front furrow width until matching occurs. Plough when conditions are more suitable.
	Tractor rear wheel too wide for furrow.	Set all mouldboards to push furrow slices further across. Use narrower wheel or wider bodies.
	Plough crabbing.	Check plough pitch – increase length of upper link. Make sure that check chains are slack. Check pitch of individual bodies – decrease.
5. Uneven ploughing	Shares worn unevenly.	Fit a new or equally worn set. Adjust bar points until extended equally.
	Pitch on bodies different.	Check clearance between underside of share points and underside of beam. Adjust until all bodies are equal.
	Mouldboard settings not the same.	Check and adjust. Adjust until all bodies are equal.
	Coulters not at the same setting.	Check vertical, lateral, longitudinal and any angling adjustments are the same for each body.
	Skim coulters taking off different amounts.	Ensure settings are same for each body.
	Skimming remaining attached to furrow slices.	Check settings and sharpness of skim coulters.
	Tractor furrow wheel sinking, causing front furrow to appear smaller or lower than the rest.	Increase front furrow width. Plough when conditions are more suitable.

Fault	Cause	Remedy
6. Furrow slices standing on edge	Depth of ploughing too great for width of furrow.	Reduce depth of ploughing. Increase furrow width. Fit mouldboard extensions. Use an alternative plough with larger width to depth ratio. Fit alternative plough bodies.
	Wrong type of body for land and speed of ploughing.	Use an alternative plough. Fit alternative plough bodies.
	Share too wide for furrow width — leaving no uncut section on which the furrow slice can pivot. *vertical cut made by coulter* *horizontal cut made by share* *uncut section necessary to enable furrow to pivot*	Fit shares with a narrower cutting edge. Cutting edge of share should be at least 50mm less than the furrow width. (Not necessarily so with a General Purpose plough.)
	Skim coulters cutting too large a skimming — preventing furrow slice from lying down.	Adjust skim coulter to reduce skimming. Adjust all skim coulters equally.
	Skimming not cut clear from the furrow slice.	Check settings and sharpness of skim
	Skimmings not being deflected far enough across the furrow — preventing furrow slice from lying down.	Adjust skim coulter settings and angle.

Fault	Cause	Remedy
	Mouldboards not scouring.	Always cover mouldboards with rust preventive when not in use.
		Clean off any rust from mouldboards.
		Replace any holed mouldboards.
		Recheck ploughing conditions.
7. Grass, weeds or rubbish showing through ploughing	Skim coulters set incorrectly.	Reset and check for wear.
	Depth of ploughing too great for width of furrow.	Reduce ploughing depth.
		Increase furrow width.
		Use an alternative plough.
		Fit alternative plough bodies.
	Wrong type of body for land and speed.	Use an alternative plough.
		Fit alternative plough bodies.
	Mouldboards not scouring.	Always cover mouldboards with rust preventive when not in use.
		Clean off any rust from mouldboards.
		Replace any holed mouldboards.
		Recheck ploughing conditions.
	No drag chains fitted.	Fit drag chains.
8. Stepped furrow wall	Rear disc coulter set too wide.	Set in as narrow as possible consistent with clean cut furrow wall. Normally 10–20mm clearance.
	Rear knife coulter stem bent – clearance too wide.	Straighten or replace.

Fault	Cause	Remedy
9. Bulging and crumbling furrow wall	Rear disc coulter set too narrow.	Set out sufficiently to achieve clean cut furrow wall.
	Rear knife coulter stem bent — clearance too narrow.	Straighten or replace.
	Loose soil conditions.	Angle top of disc coulters towards unploughed land to 'overcut' (10°–15°).
	Rear knife coulter blunt.	Replace.
	Rusty or holed mouldboards.	Always cover mouldboards with rust preventive when not in use.
		Clean off any rust from mouldboards.
10. Broken furrow wall	Plough crabbing towards unploughed land.	
	Plough operating 'nose-down' so that rear landside presses against the furrow wall too high up.	Lengthen upper link.
11. Insufficient room for tractor furrow wheel — furrow wall broken down	Furrows not thrown far enough sideways.	Increase 'throw' of mouldboards.
	Wide section or balloon tyres used.	Increase throw of mouldboard or Use narrower section tyres.

24 PLOUGHING WITH A MOUNTED REVERSIBLE MOULDBOARD PLOUGH

Your aim

is to be able to:

— check your tractor and prepare it for use with the reversible plough provided

— check the reversible plough to ensure that it is in suitable condition for ploughing and attach it to the tractor

— plan field work and then mark out the headlands and any large obstructions in the field

— plough an opening furrow (or furrows) along one side of the area to be ploughed

— set the plough to turn over level and even furrows of the same depth with both right-hand and left-hand bodies, and to bury all trash, grass, weeds, etc.

— identify and correct immediately any faults that occur whilst ploughing, and make the most effective use of power and fuel

— plough round obstructions so as to leave a minimum of unploughed land

— plough the headlands (and sidelands, if applicable) level with the main area of ploughing and with neat, level joins

— maintain the plough according to the manufacturer's instruction book, so that it works efficiently at all times

— check the plough and tractor periodically, whilst ploughing, for signs of excessive wear and for distortion or damage to parts.

△ In addition, you should aim to operate the tractor and plough safely and efficiently so that accidents are avoided and an acceptable rate of work (hectares ploughed per hour) is achieved economically. To do this, you *must* read the plough and tractor manufacturer's instruction books, follow all the recommended procedures, and understand current safety legislation relevant to tractor operation and ploughing.

You will need

— a tractor (suitable for the reversible plough)

— tractor manufacturer's instruction book

— reversible plough

— plough manufacturer's instruction book and parts list

— tool kit

— clean rags

— grease gun and oil can

— steel tape measure

— new shares and bolts

— marking out poles

— two metre (2m) straight edge

— tyre pressure gauge

— **first aid equipment which complies with the Health and Safety (First Aid) Regulations, 1981.**

You should

have received training or gained experience in:

— safe and effective operation of tractors with mounted cultivation implements

— safe preparation and maintenance of tractors for use with mounted cultivation implements, according to the manufacturer's instruction book.

Note: △ = important safety point.

1. Description of the plough

N.B.
Plough manufacturers may use different names for particular parts, so in order to avoid any confusion, the terms used in this guide are the same as those in the booklet 'Glossary of terms relating to agricultural machinery and implements' – British Standard 2468 : 1963.

(a) Identifying plough parts

Locate the parts in Fig. 1 on your own plough, and find out what each part does when the plough is working.

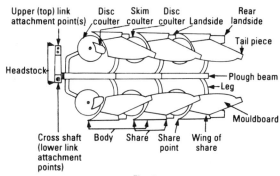

Fig. 1

To do this, read your plough manufacturer's instruction book, or get this information from someone who is already fully experienced in using that particular reversible plough, or one of a similar type.

(b) Description of parts

Understanding what each part does is important for the correct operation and maintenance of your plough.

Headstock
The complete assembly which carries the upper (or top) link and lower link attachment points as well as the turn-over mechanism for rotating the two sets of plough bodies when they are not being used.

Upper link attachment point (also called top link attachment point)
Where the upper (or top) link of the tractor is attached by a suitable pin in the recommended hole (see manufacturer's instruction book).

Cross shaft
The tractor lower links are attached to pins at each end; the cross shaft can often be adjusted laterally, but should be set centrally for reversible ploughs.

Body
The complete assembly, usually consisting of share, mouldboard, landside, tailpiece, etc. which cuts and turns over the furrow slice.

Share
This is made up of the share point (often separate) and the wing; the share point penetrates the soil and the wing makes the horizontal cut in the soil.

Mouldboard
This turns over and, in some cases, breaks up the furrow slice. It is usually concave-shaped and its length varies with the type of body (general purpose, semi-digger or digger).

Landside
This takes the side thrust of the plough when ploughing; the rear landside is usually longer than the others.

Disc coulter
This makes the vertical cut in front of the shin (the leading edge) of the mouldboard, while the wing of the share makes the horizontal cut. (Other types of coulter include 'knife' coulters and 'sword' coulters, which are lighter and have no moving parts.)

Skim coulter (if fitted, each disc coulter will have one)
This takes the corner off the furrow slice to ensure that surface trash, grass or weeds are buried completely under the furrow slice.

Tailpiece (if fitted, each mouldboard will have one)

This is an adjustable extension for helping to turn over the furrow slice and lay it against the others.

Leg

This connects the body with the plough beam.

Beam

This carries the body (attached by the leg) and the coulters; together, the beams are usually referred to as the plough **frame**.

As you will see from the plough manufacturer's instruction book and parts list, there are many other parts, but those mentioned are the main ones affecting the working and adjustment of your reversible plough.

2. General safety points

△ *Like all machines, tractors and reversible ploughs are potentially dangerous if the proper safety precautions and procedures are not adopted. When safe methods of working are combined with alertness for potential hazards the risk of accidents is minimised.*

△ *BEFORE attempting to handle a plough you should receive instruction from your supervisor/ manager, employer or instructor on how to check, maintain and operate the tractor and reversible plough safely.*

The points listed opposite apply generally to working with tractors and reversible ploughs; subsequent sections in this guide will point out specific safety points. The tractor and plough manufacturer's instruction books should be studied, and recommended safe working procedures followed at *all* times.

△ *Personal safety*

You are strongly advised to be fully immunised against **TETANUS** before working on soil-engaging components.

△ *Do not* wear loose clothing when working on or near the tractor or plough; remove your wrist watch and finger rings.

△ *Always* wear suitable safety footwear which is in good condition.

△ *Always* take care when handling heavy wheels or weights and obtain assistance if necessary.

3. Preparing a tractor for ploughing

First check that the tractor has had its regular service and then carry out the following activities and checks.

N.B.

△ *Before* dismounting from the tractor, apply the parking brake, move all controls to neutral, stop the tractor engine and leave tractor parked in first (low) gear.

△ Operate all tractor controls *only* from the driver's seat – *not* from the ground or other positions.

(a) Wheel track widths

Refer to the plough manufacturer's instruction book for the recommended rear wheel track centres and adjust them accordingly.

△ Select a firm, level site for jacking up the tractor but *do not* exceed the safe maximum load capacity of any jack or hoist used (this also applies to axle-stands and the like).

There are three methods of setting *front wheels:*

(i) *Centres of front and rear wheels in line* as in Fig. 2 below.

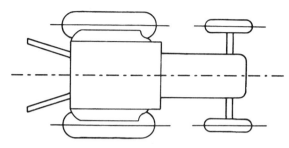

Fig. 2

(ii) *Inside walls of front and rear wheels in line.*

(iii) *Front wheels set in an intermediate position.*

N.B. The first method is recommended when ploughing *deeper* than about 200mm, because there is less wear on the inner walls of the front wheels from the furrow wall.

(b) *Tyre pressures*

Refer to the tractor manufacturer's instruction book. Pressures are usually:

front — 2.0 bar
rear   — 0.8 bar

Inflate both rear tyres to the same pressure; unless this is done, it will be difficult to set the plough to turn over even furrows.

△ Use a special pressure gauge for *water-ballasted* tyres.

(c) *Front weights*

Extra front weights are needed for most tractors to get adequate weight transfer with draught control. Refer to the plough and tractor manufacturers' instruction books.

△ Ensure that the tractor has adequate front weight for effective steering, and to keep it stable when working on hilly or sloping land.

(d) *Three-point linkage*

Refer to the tractor and plough manufacturer's instruction books for the correct category of linkage to suit the plough.

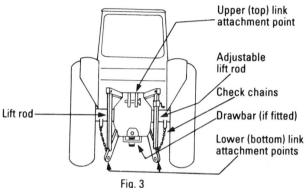

Fig. 3

The size of the tractor and plough determines which category to use. Table 1 below gives category sizes to help with identification.

Linkage	Upper link pin	Lower link pin
Category 1	19 mm diam.	22 mm diam.
Category 2	25 mm diam.	28 mm diam.
Category 3	36 mm diam.	30 mm diam.

Table 1 (to nearest mm)

Set both lift rods to the length recommended in the manufacturer's instruction book; *they must be set the same length.* Ensure that enough thread is engaged on both rods to lift the plough.

Check and adjust any other linkage parts (e.g. check chains, sway blocks, upper (top) link). If necessary, remove the drawbar (if fitted), or set it over to one side so that it will not foul the lift linkage (see manufacturer's instruction book).

(e) Hydraulic system

When raising and lowering the plough in the yard or travelling to and from the field, use position control. Use draught control during ploughing for ploughs without depth wheels. Consult manufacturer's instruction books.

4. Preparing the plough and attaching to a tractor

△ If possible, attach the plough to the tractor three-point linkage *before* carrying out checks and adjustments.

△ *Always* support the mounted plough securely (when in the raised position) *before* going underneath for any reason.

△ *Before* attachment, ensure that the plough is securely supported so that it will *not* topple over (see manufacturer's instruction book).

(a) Reversing tractor and attaching plough

Reverse the tractor squarely back to the plough's cross shaft – see Fig. 4 below.

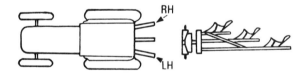

Fig. 4

First, put the tractor hydraulics in *'Position Control'* (*not* Draught Control). Then attach the three-point linkage in the following sequence:

Step 1 – Attach the *left-hand (LH) lower link* and secure it with the spring cotter pin (also called linch pin) provided.

Step 2 – Attach the *right-hand (RH) lower link*, using the levelling box on the lift rod to adjust the height of the link as necessary; fit the spring cotter pin.

Step 3 – Attach the *upper (top) link* (adjusting its length as necessary), then re-adjust it to the standard length once attached; fit the spring cotter pin.

Step 4 – Adjust the *linkage check chains* to prevent the plough from swinging onto the rear tyres during transport (but *not* so tight as to restrict the height to which the plough can be fully raised).

Step 5 – Re-adjust the *right-hand lift rod's* length to the *same* length as the left-hand lift rod, using a steel tape measure.

Note:
This attachment procedure is covered in more detail in trainee guide M.1.B.1 – 'Attaching mounted machines to the three-point linkage'.

Attach the *upper (top) link* and *lower links* to the attachment points on both tractor and plough headstock, as recommended in the manufacturer's instruction books (to get efficient weight transfer and penetration).

△ *Do not* attempt to manhandle the plough; the tractor must be reversed up to it squarely.

△ Operate the tractor hydraulics controls *only* from the tractor seat when attaching (or detaching) the plough.

△ *Before* dismounting from the tractor to attach (or detach) the plough, apply the tractor hand brake, put the hydraulic controls in the *NEUTRAL* position, stop the tractor engine and leave tractor parked in first (low) gear.

(b) Checking alignment of plough bodies

For reversible ploughing, it is essential that right-hand and left-hand sets of bodies are aligned exactly the same. If not, it will be impossible to get matched and even furrows when ploughing (see fig. 5 overleaf).

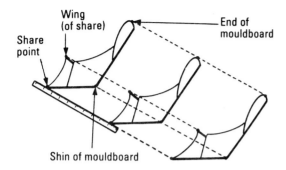

Fig. 5 Plan view of plough bodies

Carry out the following checks after fitting new shares:

Step 1 – Share points

When measured, all should be the same distance apart and in line, and should touch the ground when on a firm, level surface such as concrete (see Fig. 5).

Step 2 – Mouldboards

The distances measured between similar points (wings, shins and ends of mouldboards) should all be the same (see Fig. 5).

Note:

If any of the measurements in Steps 1 or 2 differ by more than 25mm, the plough legs, beams and frame should be checked for distortion.

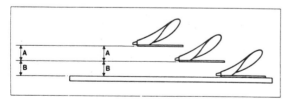

Fig. 6

Step 3 – Landsides

Remove the shares as in Fig. 6 above and place a straight edge against the landsides.

The landsides (bodies) should be parallel with each other to within 5mm in 1 m. A and B in Fig. 6 should equal the furrow width of the plough.

(c) Basic plough settings

Whilst still in the yard or workshop, carry out the following checks of the plough:

(i) *Body Pitch* (see Fig. 7 below)

Measure distance A from share point to plough beam with new shares fitted.

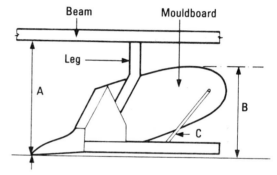

Share point
A = Underbeam clearance.
B = Mouldboard height.
C = Mouldboard stay adjustment.

Fig. 7

Distance A should be the same for each body of both sets of bodies, and as recommended in the plough manufacturer's instruction book (which should be consulted as to the method of adjustment). The height B of the mouldboards in Fig. 7 should also be the same for *all* bodies.

(ii) *Cross shaft and plough frame*

Refer to the plough manufacturer's instruction book for details and method of adjustment, depending on the furrow width and number of bodies, etc.

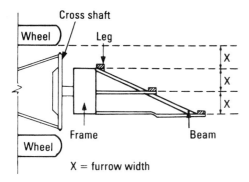

Fig. 8

The *cross shaft* should be set centrally; if it is not, the right-hand and left-hand front furrow widths will be unequal during ploughing.

The *plough frame* should be set correctly (see manufacturer's instruction book) for the number of bodies, furrow width of the bodies, type of turn-over, etc. so that the front furrow width is the same as the other bodies' furrow widths (see Fig. 8 above).

(iii) *Plough turn-over mechanism* (also called the reversing or indexing mechanism). *First* make sure that there is sufficient clearance for the bodies to rotate without striking the ground and then check that this mechanism works correctly (see manufacturer's instruction book).

△ *Do not* operate the plough turn-over control when standing beside the plough — *only* do so from the tractor seat.

△ *Before* reversing the plough bodies, make sure that everyone and everything in the vicinity of the plough is well clear.

△ When carrying out checks and adjustments ensure that the plough bodies will *not* turn over accidentally.

(iv) *Coulters*

Check all coulters (disc, knife, skim) for wear; replace or sharpen them. Set the coulters for normal ploughing conditions (see Section 7 of this guide).

△ Take care not to drop *disc coulters* onto your feet when adjusting them.

(v) *Lubrication*

The main points are:

— plough turn-over and locking mechanisms

— disc coulter bearings

(see plough manufacturer's instruction book for details).

5. Marking out

(a) *Measuring headland width*

Allow at least *twice* the combined length of the tractor and plough for turning *or* enough space to turn the tractor easily once the plough is fully raised at the end of the furrow.

△ Leave plenty of room for error when the headland is beside a hazard such as a deep ditch or stream.

(b) *Marking out headlands*

Either

follow stubble lines of combine wheelings where possible,

or

use a length of rope with the help of an assistant,

or

drive to marking poles previously placed to mark out the headlands.

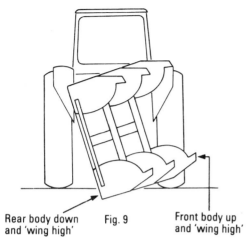

Rear body down and 'wing high' Fig. 9 Front body up and 'wing high'

Set the plough as follows:

Step 1 – Lengthen the *upper (top) link* to lower the rear body.

Step 2 – Adjust the *plough frame* to get the plough bodies 'wing high' (as in Fig. 9).

Step 3 – Set the *rear disc coulter's* cutting edge down level with the underside of the share.

Result – the rear body will plough a shallow furrow of about 75mm depth to mark out headlands (and ridges, if required).

Turn the marking furrow *towards* the centre of the field to get a tidy end to the furrows, except when soil is hard. If the plough has difficulty in penetrating hard soil, turn marking furrow *away* from centre of the field (penetration is then assisted and a cushion of soil is left to protect the shares).

△ Identify and mark out clearly any hidden depressions or holes which cannot be seen clearly in the field.

After completing the marking out, *remember* to:

— equalise the *lift rod lengths* if the right-hand one was adjusted (by means of the levelling lift rod)

— re-set the *coulter* to the correct position for normal ploughing.

6. Opening up

(a) Two basic methods

(i) *Starting at the side of the field*

This is used where *no* sidelands are needed, because both headlands can be ploughed without having to run over the ploughed land (e.g. when ploughing half a field).

(ii) *Starting along an opening ridge* (see Fig. 10 below).

This is usually done when the headlands and sidelands are ploughed together after completing the main area of ploughing (see Section 11 of this guide).

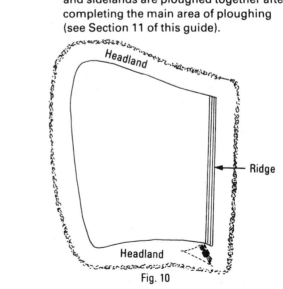

Fig. 10

(b) Ploughing the opening ridge

Follow the procedure described in trainee

guide M.6.A.1 'Ploughing with a conventional mounted plough' – for arable or grassland openings.

Note:
It is better to start the ploughing with a straight opening furrow, rather than follow the bends and curves of the field boundary (see Fig. 10). The finished ploughing looks better, but, more importantly, it is easier to achieve good standards of ploughing.

7. Setting the plough

The requirements of good quality commercial ploughing are:
— levelness of ploughed surface
— even depth of ploughing
— trash/weed/grass all buried completely
— neat ends of furrows ('ins' and 'outs')
— straight furrows
— time and cost of ploughing kept as low as possible.

To achieve these standards of ploughing, a systematic approach to setting the reversible plough is needed.

△ Operate the tractor hydraulic lift controls *only* from the tractor seat.

△ Operate the plough turn-over mechanism *only* from the driver's seat, *after* first ensuring that eveyone and everything is well clear and that the plough is fully raised.

Reversible plough setting procedure:
— *Read the manufacturer's instruction book for guidance on how to make the various adjustments to the plough.*

Set the right-hand bodies correctly on the first run so that they turn over level furrows of even depth, then adjust the *left-hand bodies* on the next run to turn over furrows which match.

The KEY POINT is:
— *make one adjustment at a time and check the result BEFORE making any further adjustments.*

△ *Always* lower the plough gently to the ground first *before* making any checks or adjustments (*or* support the plough securely in the *raised* position before going underneath).

△ When working on the plough in the *raised* position (see manufacturer's instruction book), *always* make sure that the plough bodies cannot accidentally turn over.

△ Take care to avoid being cut by sharp edges of plough parts, such as disc coulters, worn share points, worn tailpieces.

(a) Right-hand bodies

Step 1 – adjust the *pitch* of the plough frame by means of the upper (top) link to get the frame parallel to the ground, seen from the side with the plough in work – see Fig. 11 below.

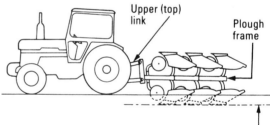

Fig. 11 – Pitch   Furrow bottom

When the plough pitch is correct, the heel of the rear landside should leave a light mark (not more than about 5mm deep) on the furrow bottom (with most ploughs).

Step 2 – adjust the plough frame *level* so that the plough legs are upright (right-angled to the ground level) when seen from behind with the plough in work – see Fig. 12.

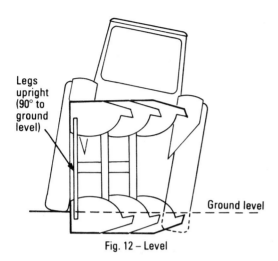

Fig. 12 – Level

Do not use levelling box on right hand lower link; use the independent levelling adjuster on the headstock.

Step 3 – Adjust the *front furrow width* until equal with the other plough bodies' furrow widths, after measuring with a tape measure. Also check visually that top link is in a straight line.

If a correction of *more* than about 50mm is required, check the tractor and plough basic settings; *do not* use the front furrow width adjustment on the plough. Check whether the rear wheel track width and plough frame-to-headstock positions are correct for the type and number of bodies, furrow width, etc. (see plough manufacturer's instruction book).

Step 4 – Adjust the *depth lever* on the tractor hydraulic system (or *depth wheel*, where used) to get the front body working at the required depth (e.g. 200mm).

Step 5 – Adjust the *coulters* to suit the conditions.

(i) *Preliminary settings for disc coulters* should be (see Fig. 13):

— *hub* of disc over point of share, and not fouling ground

— *disc* cutting up to 100mm into land, depending on conditions (hub must *not* foul ground)

— *cutting edge* of disc 12mm clear of side of share (see (a) in Fig. 13 below)

— *angle* of disc normally vertical.

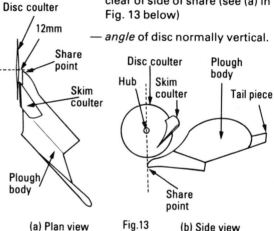

(a) Plan view　　Fig.13　　(b) Side view

(ii) *Skim coulters* should be set (see Fig. 13 above):

— just deep enough (about 50mm) to cut a continuous skimming of top soil from furrow and bury all trash

— with *tip* as close as possible to disc without actually touching it (see (a) in Fig. 13), and set below disc rivets

— with *heel* from 3mm to 12mm clear of disc, depending on ploughing conditions (see (a) in Fig. 13).

△ *Do not* place your feet immediately beneath *disc coulters* when adjusting them (in case they slip and fall).

(b) Left-hand bodies

Step 1 – Check whether the plough frame is set *level* (legs upright). If the left-hand bodies can be levelled independently of the right-hand side then they should be adjusted level if necessary.

If no independent means of adjustment for the left-hand bodies is provided then consult the plough manufacturer's instruction book.

Do not use the levelling box on the lower link to level the left-hand bodies, unless instructed to do so, as this will upset the level of the right-hand side as well.

Step 2 – Set the *depth control* the same as for the right-hand side (or set the *depth wheel* the same, if provided).

Step 3 – Adjust all left-hand *coulters* the same as the right-hand ones (see Fig. 13).

Step 4 – Drive the tractor at the *same speed* when ploughing with the left-hand bodies as you used for right-hand ploughing (if you do not, it will be impossible to match right and left-hand furrows correctly).

CAUTION: Do *not* adjust any of the following:

— *front furrow width* (any adjustment will affect right-hand side, making it too wide or narrow)

— *plough pitch setting* by means of upper link (this, too, will affect right-hand side)

— *adjustable right-hand lift rod* (unless plough manufacturer's book says so).

Note:
These left-hand settings should be found to be correct when the right-hand side has first been set correctly; if not, there is a fault (distortion) or the plough bodies have not been correctly set (as in Sections 4 (b) and (c) of this guide).

It is essential that the tractor also be prepared correctly (wheel widths correct, tyre pressures equal, etc.) otherwise matched ploughing from right-hand and left-hand sets of bodies is impossible to achieve without continuous re-adjustment of the plough.

(c) Attention points when setting reversible ploughs

(i)   Field settings

Once the initial settings of 'pitch', 'level' and 'front furrow width' have been made, *a beginner should carry out only one adjustment at a time* to correct faults. Check each result *before* making another adjustment to the plough.

(ii) *Matching the right-hand and left-hand furrows*

Often a complete bout must be ploughed before the overall effect of any adjustment can be seen. When a setting on the right-hand bodies is altered, the left-hand bodies are usually affected as well, so that the effect of the adjustment cannot be judged until the left-hand bodies follow the right-hand furrows just ploughed on a subsequent bout.

(iii) *Visual appearance of adjacent furrows*

Do not be confused by the difference in the 'look' of the right-hand and left-hand furrows lying adjacent to each other. This is caused by the 'grain' of the soil being different because the direction of ploughing has been reversed. This visual difference should *not* be allowed to mask real faults that are there (e.g. uneven and poorly-matching furrows, incorrect front furrow width, etc.)

(iv) *Upper (top) link setting*

Set the upper link as long as possible, providing the plough penetrates satisfactorily and a *light pressure* is maintained by the *rear landside* on the furrow bottom. If the upper link is set *too short*, the plough will run *'nose down'* with a considerable increase in draught.

(v) *Visual check of 'pitch'*

Once the plough has been set 'level' (legs upright), it is easy to check whether the pitch of the plough frame is correct (see Fig. 14).

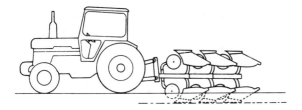

Fig. 14

As well as checking the mark left by the rear landside, look at the *height of the soil coming off the front and rear mouldboards*. If the height is greater on the front body than on the rear, the front body is working deeper (*'nose down'*); similarly, for the rear body when working deeper (*'tail down'*) the soil will be higher than on the front body.

Adjust as follows:

'nose down' – *lengthen* upper link

'tail down' – *shorten* upper link

until the height of the soil coming off the front and rear mould boards is the same.

(vi) *Front furrow width*

If adjustments are made to the frame 'level' setting, the front furrow width may have been altered as well by a few mm. Check and, if necessary, re-adjust the front furrow width after any 'level' adjustments.

(vii) *Upper link alignment*

From the driver's seat, the upper (top) link should be *parallel* with the line of draught of the plough (see Fig. 15 below). However, a slight inclination of the upper link towards the furrows will usually have no adverse effect.

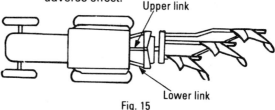

Fig. 15

The distance between each lower link and the tractor wheel should be *the same*, and the check chains should hang freely by the same amount.

If not, check for incorrect basic settings of the tractor and plough (Sections 3 and 4), and check the plough 'pitch', 'level' and 'front furrow width' settings. The front furrow width setting is often the cause of this fault.

(viii) *Tractor 'crabbing'*

If the tractor front wheels 'crab' away from the land, requiring the wheels to be steered into the furrow wall, the plough is *not in balance*. The top link will also point to one side or the other.

Check and adjust basic settings and field settings as necessary; otherwise, fuel consumption may well be increased very considerably (up to 30 per cent more).

Check for any of the following faults:

— Is the plough frame set nose down? Wing down?

— Is the front furrow width set too wide?

— Are the front wheels set too wide?

— Are the rear wheels set too narrow?

— Are the check chains (one or both) set too tight?

— Is the headstock incorrectly positioned in relation to the main plough frame?

Correct any faults found according to the plough/tractor manufacturer's instructions.

(ix) *Depth and speed settings*

Unless the same depth and speed are used for *both* right-hand and left-hand ploughing it will be impossible to match up adjacent furrows. (Beginners often drive slower when ploughing with left-hand bodies, and do not set the depth control the same as for right-hand ploughing.)

8. Driving the tractor

(a) Turning on headlands

Lift the plough out of the soil; *turn away* from ploughing when doing a three-point turn on the headland, then reverse the tractor into line with the open furrows of the previous run. Alternatively, a loop turn can be used, if space permits.

△ Take care when turning that the plough does not strike anyone or anything as it swings round in an arc behind the tractor.

△ *Take special care* when operating a reversible plough on slopes or hilly land, especially when turning with the plough in *raised* position.

△ When operating on a slope *always* turn slowly, and whenever possible *turn up* the slope, using a three-point turn. Ensure that there is sufficient ballast weight attached to the *front* of the tractor (see HSE leaflet 'Safe tractor driving on slopes').

(b)'Ins'

Lower the plough into work as the *rear wheels* of the tractor cross the *headland marking furrow*. To get the plough to the required depth as quickly as possible, push the depth control lever *below* the selected position on the quadrant, then raise it to its predetermined position.

(c) 'Outs'

To get even depth at the end of the furrow, push the depth control lever *down* slightly as the tractor front wheels rise over the headland marking furrow. Then raise the plough out of work as the *rear wheels* cross the *headland marking furrow*.

N.B.

This procedure and that above for 'Ins' will result in neat ends to the furrows where they meet the headland marking furrow.

(d) Lowering the plough into work

Always lower the plough gently, particularly when soil is hard; otherwise, share points or wings may break from the shock of hitting the hard ground.

(e) Reversing the plough

△ Make sure that nobody and nothing will be hit by the plough when the plough turn-over mechanism is operated.

Do not operate turn-over mechanism with the tractor rear wheels in a hollow, or the plough may strike the ground.

(f) Tractor hydraulics

Use the slowest 'response' setting possible, whilst still maintaining an even depth of ploughing. This allows for maximum transfer of weight from plough to tractor (when there is sufficient weight added to the front of the tractor).

If the soil varies considerably from heavy to light along the field, adjust the *'draught control' lever* slightly to maintain the required working depth. If the plough 'bobs' up and down, move the 'response' control towards *slow*.

For other adjustments to the hydraulics and linkage, refer to the manufacturer's instruction books for tractor and plough.

(g) Differential lock

When the going is heavy and sticky, *use the differential lock all the time*, except when turning. The diff lock should *not* be used just when wheel-slip becomes excessive.

△ *Always* disengage the diff lock at the headlands before beginning to turn the tractor.

△ If one wheel is spinning do not engage the diff lock without first depressing the transmission clutch.

(h) Share wear

Share the ploughing as equally as possible between the right-hand and left-hand bodies; otherwise, the shares will wear more quickly on one side. This will make matched ploughing difficult to achieve. Often, only the right-hand bodies are used for marking the headlands, ploughing odd corners and ploughing headlands – *this should be avoided*.

△ Keep well clear of ditches, dykes, streams and the like when turning on or ploughing the headlands.

△ Take care to avoid hidden depressions masked by undergrowth which may be in the field being ploughed.

△ *When travelling on the road* ensure that the tractor brakes are balanced, and that the pedals are latched together securely. Always 'trip' the plough before travelling, so that the trip mechanism is loose, and therefore safe.

9. Correcting faults

(a) Requirements

When the tractor and reversible plough have been prepared and set correctly, the ploughing can proceed properly with a minimum of problems.

The tractor should run smoothly, without the engine labouring, or excessive wheelspin.

The tractor should virtually steer itself and the upper (top) link should be in line with the line of pull when the tractor and plough are 'in balance' (not working against each other).

Apart from the quality of the ploughing, there are two other indications of whether the tractor and plough are 'in balance' with each other:

(i) if the *tractor front wheels* are running or pushing at an angle (instead of being in line with the rear wheels), the tractor and plough are *not* set correctly in relation to each other

(ii) if the *upper (top) link* is pointing to the left or right of centre while ploughing, this is also a sign that the tractor and plough are *not* 'in balance' (see Fig. 16 below).

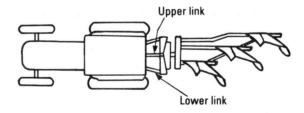

Fig. 16

When the tractor and plough are *not* 'in balance':

— there is increased draught and wear and tear
— more time and fuel are used

— operator's job is made more difficult
— standards of ploughing (quality) will be lower.

(b) Fault-finding and correcting

While experience of ploughing is being gained, it helps to follow a systematic approach to fault-finding and faults correction as described below:

1. Recognise the *fault* in the first place.
2. Identify the *cause* or causes.
3. Carry out any necessary *remedial action* promptly to correct the fault.

The *Faults Chart* for Reversible Ploughing (supplied separately) should help you identify and correct faults.

As far as a beginner is concerned, the key point is to *correct one fault at a time, and make only ONE adjustment at a time.* Then plough on a bit further and check the effect of the altered adjustment (is the fault better or worse?) *before* making any further adjustments.

10. Ploughing round obstacles

(a) Small obstacles (e.g. trees, pylons, stacks and the like)

When the obstacle prevents further uninterrupted ploughing, raise the plough out of work at A (see Fig. 17 below) about 20 metres from the obstacle. Turn over the plough bodies and reverse the tractor back to B.

Lower the plough into the ground at B and plough back to A so that the ploughing joins neatly.

Lift the plough and turn over the plough bodies again; then drive round the obstacle.

Reverse the tractor back to point C on the other side of the obstacle; lower the plough into the ground with the tractor wheels aligned with the open furrow.

Continue ploughing on up the field to the headland. On the return run, repeat the same procedure for the other side of the obstacle.

Continue in this manner until the obstacle no longer obstructs the tractor and plough; then resume normal ploughing.

(b) Large obstacles (e.g. buildings, ponds, spinneys and the like)

Mark out a headland around the obstacle wide enough on which to manoeuvre the tractor and plough safely (i.e. about *twice the combined length* of the tractor and reversible plough, or wider).

Plough out the headland in the normal manner, when the surrounding ploughing has been completed.

△ *Take care* when ploughing near and around obstacles which are potential hazards, such as ditches or streams, steep slopes and banks, or, *especially*, overhead wires on pylons.

11. Ploughing the headlands

(a) General points

(i) Plough the headland IN (furrows turned *away* from the hedge) one year, and OUT (furrows turned *towards* the hedge) the next.

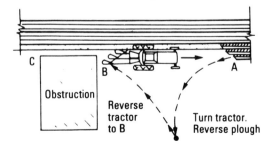

Fig. 17

(ii) Use *both* sets of plough bodies (right-hand and left-hand) the same amount so that both sets of shares wear equally.

(iii) Plough any small areas first to reduce to a minimum any running over ploughed land by the tractor (causing 'wheelings').

(iv) Whenever possible, plough the headland *nearest the gate* last of all to minimise 'wheelings'.

(b) Headland ploughing procedure

(i) *Ploughing the headland OUT*

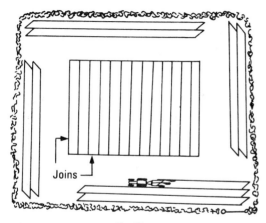

Fig. 18

Start at either end of the headland as near the hedge as possible – see Fig. 18 above. Plough around the main area of ploughing until the headland is ploughed out.

Make sure that the 'joins' are neat and level with the rest of the ploughing.

(ii) *Ploughing the headland IN*

Start at either end of the headland along the main area of ploughing, so as to turn over furrows, making a 'join' with the previously ploughed land.

Plough around the main area of ploughing until the headland is completed (see Fig. 19 below). Plough as close as possible to the hedge (or field boundary) when finishing the headland.

△ *Take care* when ploughing near the field boundary; look out for potential hazards, such as overhanging branches, ditches or steep banks.

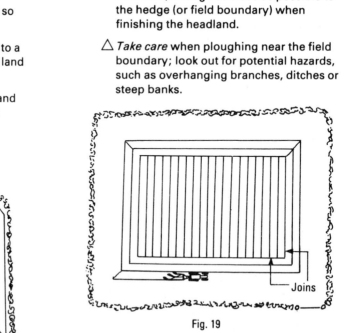

Fig. 19

12. Maintaining and checking the plough

(a) Routine checks

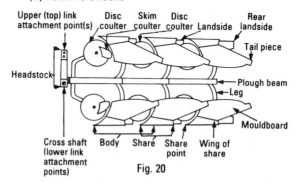

Fig. 20

(i) Check the plough periodically (daily when new and thereafter at least once a week) for loose nuts or bolts, damaged or worn parts, as well as distorted or misaligned plough parts.

(ii) Check that the plough turn-over and locking mechanism and other working parts (e.g. depth wheels) are correctly adjusted and functioning correctly.

Refer to the plough manufacturer's instruction book for instructions and maintenance details.

△ *Always ensure that the plough cannot drop, topple over, or turn over accidentally when checking and working on it.*

(b) Routine lubrication

Refer to the plough manufacturer's instruction book.

Generally, each lubrication point will require attention *once a day*, but some points (e.g. disc coulter bearings) may require lubrication more than once daily. Some points will require lubrication with oil; others will require grease.

(c) Soil-wearing parts

Check all soil-wearing parts for wear and replace them when excessively worn. Parts to check include plough shares, wings, mouldboards, shins, landside heels and tailpieces as well as coulters and skims.

Replacement of *plough shares* at intervals during the ploughing will be necessary as the work proceeds. Worn shares affect penetration of the plough, particularly in hard, dry soil conditions when wear is accelerated as well, and should be replaced as soon as they are worn too much (see opposite).

Replace worn shares:

either

— when there is *insufficient clearance* for normal working (see Fig. 21)

or

— when *penetration is affected* by hard condition

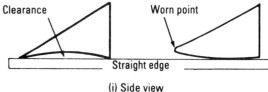

(i) Side view

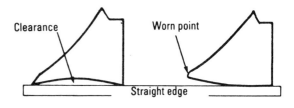

(ii) Top view

Fig. 21

Partly-worn shares can be re-used later in softer soil conditions.

Check that shares are fitted correctly after cleaning the frog (the part to which the share is attached) to allow proper seating of the share. Tighten the fastening nuts/bolts securely.

(d) Storage

(i) *Overnight storage*

Protect the soil-wearing parts, especially mouldboards, from rust by applying oil or grease on all shiny surfaces *each night* (or when ploughing is finished).

(ii) *End of season storage*

Coat all soil-wearing parts (shares, mouldboards, disc coulters, etc.) with an anti-rust preparation which will protect them from rust for long periods. Oil or grease are *not* suitable.

If possible, keep the plough under cover during the period it is in storage.

△ *Make sure that the reversible plough is safely and securely supported during storage so that there is no chance of its toppling over onto anyone (see manufacturer's instructions book for details).*

You should understand

the importance of:

△ — correct and safe working and operating procedures when using and maintaining a reversible plough

— preparing both the tractor and reversible plough according to the manufacturer's instructions *before* ploughing starts

— adjusting the plough in the field to produce acceptable quality ploughing economically, with the minimum of wear and tear on either tractor or plough

— promptly recognising and correcting faults in ploughing, or in the plough and tractor

— maintaining and checking the tractor and plough at regular intervals (usually daily).

If you do not understand all these points, ask your instructor or manager/supervisor for further instruction or advice, particularly about the safety points and procedures.

△ Then, tick off the activities in the list below which you are sure that you can carry out *safely*. If you are still in any doubt, ask your employer/manager/instructor to explain further – *your life or someone else's could depend on it!*

Preparing tractor for ploughing .... ☐

Preparing plough .... ☐

Attaching plough to tractor .... ☐

Operating tractor when ploughing .... ☐

Transporting plough to and from field .... ☐

Operating plough (especially on slopes) ..... ☐

Carrying out checks and maintenance .... ☐

Further experience

You will become proficient only by practice. Try to get experience of reversible ploughing as soon as you can by taking every opportunity offered by your manager or employer to follow up the training you have received.

Remember that the Board and your county agricultural college offer a wide range of courses in practical skills, together with associated knowledge, which will help you understand more about the work your are doing.

25 PRE-SEASON PREPARATION OF SEED DRILLS

Your aim

is to be able to:

— remove drill from store

— clean, lubricate and refit parts removed for storage

— match correct size tractor to drill

— carry out machine test to check all parts function correctly

— identify malfunction, diagnose faults and correct as necessary

— set drill seed and fertiliser application rates

— set drill to function correctly in varying soil conditions

— carry out stationary calibration on seed and fertiliser application rates

— prepare drill for road transport or load onto transporter

— select and wear the appropriate protective clothing when handling treated seed

— carry out moving calibration of seed and fertiliser.

You will need

— your farm's drill (which should have been maintained and stored correctly at the end of last season)

— tractor of suitable size

— manufacturer's instruction book and parts list

— calibration trays (if supplied with drill) or sheet

— hand tools and calibration charts including any tools provided with the drill

— set of accurate scales up to 5 kg

— pocket steel tape measure

— grease gun, oil can and correct grades of lubricant

— air line incorporating a tyre pressure gauge

— length of wood just over 1 metre in length ⎫
— 2 x 150 mm nails   ⎬ to make line for field calibration
— ball of baler twine   ⎭

— flat area of concrete, well lit and preferably under cover

— de-greasing fluid, drip tray, brushes and rags

— sufficient seed and fertiliser for calibration (50 kg bag)

— protective clothing as required by the Health and Safety (Agriculture) (Poisonous Substances) Regulations 1975

— First aid equipment which complies with the Health and Safety (First Aid) Regulations, 1981.

You may need

— cage wheels

— front and rear tractor weights

— drill weights where applicable

— jack and axle stands

— sheet or containers to catch materials when calibrating

— brush, shovel

— transporter or transport wheels.

You should

have received training or gained experience in:
— the safe handling of chemicals
— Safe lifting and carrying
  (trainee guide COM.1.1)
— Altering wheel track widths
  (trainee guide M.1.C.4)
— Attaching mounted machines to three point linkage (trainee guide M.1.B.1)
— Hitching trailed machines
  (trainee guide M.1.B.2)

Note: △ = important safety point.

1. Introduction

This guide covers work that will need to be done assuming that the drill was maintained and stored correctly at the end of last season. If this was not done then further work may be necessary now.

Read the manufacturers' instruction books for both tractor and drill before starting work and refer to them for details of settings.

2. Preparing the tractor

Match the tractor to the drill considering:

— category of three point linkage required
— hydraulic couplings, valves and pipes
— drawbar height
— power requirement bearing in mind slope of field and ground conditions. For trailed drills, a useful guide is 3 kW (4 hp) per row.

Follow drill manufacturer's recommendations on matching.

Check that brakes are correctly adjusted.

Ensure that the tractor is large enough to form a stable combination with the drill (particularly mounted drills).

△ Check tyres of drill and tractor for:

— correct pressures
— dust caps
— condition.

Report any faults.

Fit tractor front end weights and rear wheel weights if necessary.

Remember that with mounted machines sufficient weight is required to keep front wheels on the ground and maintain effective steering when hopper is fully loaded.

To increase stability set the tractor wheels as wide as practical (trainee guide M.1.C.4).

Fit automatic coupling frame if drill is fitted with automatic hitch.

3. Removing drill from store

Jack up drill and remove blocks or axle stands.

Attach drill to tractor.

a) Mounted machines

— attach three point linkage or automatic coupling
— adjust right hand lower link until drill is level
— adjust top link until drill is at the angle recommended in the manufacturer's instruction book.

N.B. It is particularly important that cultivator drills should be level from fore to aft and from side to side, otherwise uneven sowing depth and irregular emergence will result.

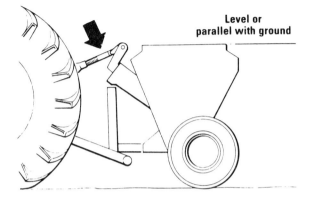

Level or parallel with ground

Set top link to manufacturer's recommendations.

b) Trailed machines

If the drill drawbar is in the transport position support its weight, remove the retaining pin and carefully lower.

Refit pin in working position and secure.

Set drawbar hitch plate in correct position to give correct working height and correct drill attitude. (See manufacturer's instruction book.) It is usually higher for tine coulters than disc coulters.

Adjust tractor drawbar to this height.

⚠ Hitch to tractor using a drawbar pin of the correct size and in good condition and a lynch pin to secure it. Clean hydraulic couplings carefully before connecting them. Place all caps for the couplings in the tractor tool box when not in use. Ensure that tractor is equipped with correct hydraulic services (most drills require a single or double acting spool valve).

Ensure that hydraulic pipes will not rub on the drill causing wear.

Check all hydraulic pipes for damage or wear.

Replace any that show signs of deterioration.

Wind up drill jacks and place in stowage positions and secure them.

Carefully operate the hydraulic lever to check free movement of all parts, after making sure no-one is near the drill.

Ensure coulters are in raised position before drawing forward.

Wear the prescribed protective clothing if the drill has been used with treated seed.

Brush off any loose dust, dirt or bird droppings that may have gathered during storage.

Clean off any rust preventative that was applied prior to storage, using a suitable solvent.

Consult manufacturer's instruction book as some nylon type feed rollers can absorb these materials, swell and then seize in their housings.

4. Checking drill

Check that any faults found during the end of season maintenance have been rectified.

Recheck drill for parts that may have deteriorated during storage.

Check the condition and location of the dividing plate in the hoppers (combine drills).

N.B. This can be set in various positions to vary size of seed fertiliser hoppers. Select best combination to minimise time spent in loading.

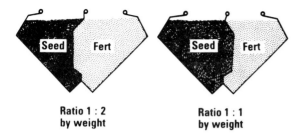

Ratio 1 : 2 by weight

Ratio 1 : 1 by weight

Select best position for hopper dividing plate depending on ratio of seed to fertiliser.

Ensure that the fertiliser hopper sieve is in place. Never use the drill without the sieve as large lumps of fertiliser can damage the metering mechanisms.

Remove all down spouts and coulter tubes if appropriate and check for cracks.

Replace as necessary.

Ensure that telescopic coulter tubes can telescope freely.

Re-assemble any parts dismantled for storage.

Adjust tension on any chains and pulleys, taking care there are no high spots where chains run on cast sprockets as this may snap the chain.

Check condition of rubber tubes and replace as necessary.

N.B. When re-assembling, plunging in warm water makes them more pliable. A few drops of washing up liquid also helps lubricate them.

Allow sufficient time for them to dry before drilling small seeds as they may stick and cause a blockage.

Check spring loaded flaps in feed mechanism to ensure that they can return to original position (especially important with fertilisers).

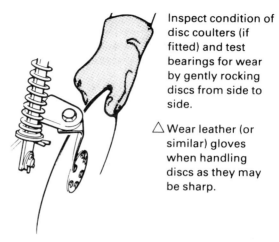

Inspect condition of disc coulters (if fitted) and test bearings for wear by gently rocking discs from side to side.

△ Wear leather (or similar) gloves when handling discs as they may be sharp.

Adjust shoe or scraper correctly. (Normally so that it is as close to the disc as possible without actually impeding its motion.)

(See manufacturer's instruction book.)

Check leading edge of Suffolk type coulters and points of hoe coulters; replace or reverse as necessary.

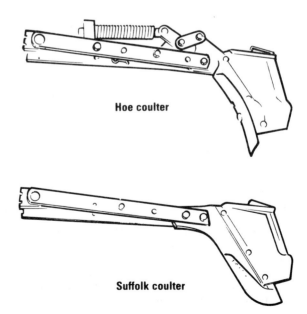

Hoe coulter

Suffolk coulter

Ensure that all nuts and bolts are tight.

Check wheel nuts and any other nuts and bolts that require a specific torque setting.

Check that vermin have not got inside hoppers and left residues or caused damage, particularly by gnawing at metering units.

Check adjustment of engage/disengage mechanism.

Rotate feed rollers with calibration handle to ensure free movement.

Do not force any seized parts, carefully free them with penetrating oil before repeating.

Check condition of points on track eradicators (if fitted) and reverse or replace as necessary.

Check condition and adjustment of covering harrow (if fitted).

Ensure that spares of parts that normally wear are available for replacement during the season or that they are ordered in time.

△ Ensure that all guards are in place and fully effective. Notify any deficiency to your manager.

5. Lubrication

Grease all nipples (using manufacturer's instruction book).

Wipe outside of bearings before and after greasing.

Oil chains and sprockets except those in contact with the soil (see manufacturer's instruction book).

Check gearbox oil levels and top up with correct grade of oil or change oil if necessary.

6. Setting row spacing

Use a setting board if provided with the drill and follow manufacturer's instructions. If no setting board is provided, set row spacing as follows:

— find centre mark of drill

— transport drill to an open area of ground where a trial run of 2-3 m can be made. Turf is best for disc coulters, cultivated soil for Suffolk or hoe type coulters.

— drive forward slowly and lower coulters into the ground

— drive for 2-3 m and stop

— find centre position and measure on the ground the distance between each impression made by the coulters.

This will reveal any inaccuracy in row spacing caused by sideways movement in the coulter linkage.

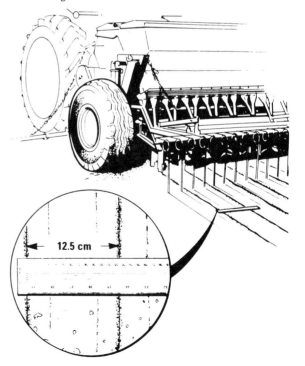

Correct row spacing by:

— slackening clamping device

— sliding coulter block along frame

— retightening clamp.

Repeat test run until correct setting is achieved.

Always work from the centre outwards.

If you do alter coulter spacings be sure to adjust the push rods also.

If a wider row spacing is required other than that for which the drill is set, use the blanking plates to shut off the feed to the selected coulters.

It is often necessary to increase pressure on coulters folowing in tractor wheel tracks.

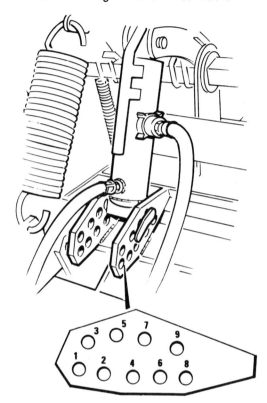

7. Setting coulter depth

Set the coulter depth (using the manufacturer's instruction book).

Normally coulter depth should be approximately the same as drilling depth. However speed of drilling has an effect, i.e. the faster you travel the shallower the penetration.

Consequently, check actual drilling depth in the field and readjust if necessary.

Adjust coulter spring pressures if necessary.

8. Setting bout markers

Markers should be set for either wheel marking or centre line marking.

a) Wheel marking

Use the following formula to work out the position of the marker.

$$\text{Distance from the last coulter of the drill to the marker disc} = \tfrac{1}{2} \text{ working width of the drill} - \tfrac{1}{2} \text{ tractor wheel track width} + 1 \text{ row width}$$

Where working width of the drill = distance between coulters x number of coulters.

b) Centre line marking

| Distance from the last coulter of the drill to the marker disc | = | ½ working width of drill | + | ½ row width |

Example (see illustration)

Number of coulters = 20
Drill working width = 20 × 175 = 3500 mm (A)
Tractor wheel track = 1525 mm (B)
Row spacing = 175 mm (C)

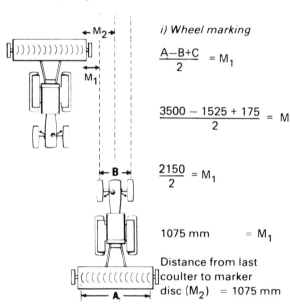

i) Wheel marking

$$\frac{A - B + C}{2} = M_1$$

$$\frac{3500 - 1525 + 175}{2} = M_1$$

$$\frac{2150}{2} = M_1$$

$$1075 \text{ mm} = M_1$$

Distance from last coulter to marker disc (M_2) = 1075 mm

ii) Centre line marking

$$\frac{A + C}{2} = M_2$$

$$\frac{3500 + 175}{2} = M_2$$

$$1837.5 = M_2$$

Distance from last coulter to marker disc (M_2) = 1837.5 mm

9. Setting tramlining equipment

Set tramliner to blank off rows at the normal tractor wheel track width used on the farm if possible.

This will save altering the wheel track for subsequent operations.

If blanking off a single row for each wheel does not give a wide enough tramline (e.g. if 35 cm (14 in) tyres are in use) then move the coulters either side of the tramline out up to 5 cm to give increased width.

Lubricate chains and sprockets.

Adjust tension of chains (if appropriate); if they are too slack grain will not be fully blanked off, if too tight they may cause binding in operation.

On manually operated systems run ropes into rear of tractor cab ensuring that they are not fouling any moving parts.

⚠ On mechanically operated systems run any hydraulic couplings and electrical cables to the tractor and see that they are secured where they will not foul moving parts.

Ensure that any controls are secured in a safe position and clearly marked.

⚠ Ensure that hydraulic levers are in the neutral position before making the coupling.

Place caps for hydraulic couplings in the tractor tool box when not in use.

⚠ Ensure that the power is not connected or is switched off before connecting any electrical cables to the control box.

⚠ Connect up electrical cables and switch on ensuring that no one is near the mechanism.

Set the system for the appropriate bout number.

Carefully test the system to ensure free movement and correct sequence of operation.

On systems which mechanically or electronically count the number of times the drill is raised, ensure that the 'clutch' or 'trip' mechanism used when raising the drill, other than at the end of a run, functions correctly.

With manual systems where bout counting relies on memory, it is useful to record bouts e.g. a string with clothes pegs suspended in the cab; slide one peg along at the end of each bout.

Consult the manufacturer's instruction book if a fault develops.

10. Setting drill for seed size

When drilling small seeds, insert small seed bottles or grass hoppers for grass seed (see manufacturer's instruction book).

Set hopper aperture to the correct setting.

Adjust flaps or shutters on the metering mechanism to recommended opening.

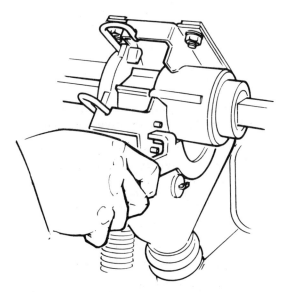

Change feed roller or metering devices.

11. Setting seed and fertiliser application rates

Seed rates and fertiliser rates are achieved by the selection of the correct gear and quadrant setting, the correct cassette setting or the correct gearbox setting.

Set the drill to the required application rates using the manufacturer's instruction book.

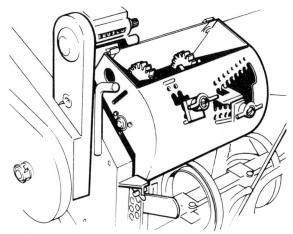

Select the correct gear and quadrant setting

Ensure that the drive belt to the hectaremeter/ acremeter is on the correct pulley.

12. Calibrating the drill

a) Stationary calibration

Follow the procedure outlined below unless the manufacturer's instruction book specifies an alternative.

1. Disengage drive to fertiliser side if possible, if not then leave hopper empty.

2. Fill grain hopper with enough seed to drill about 0.2 hectares ($\frac{1}{2}$ acre).

3. Spread grain evenly across hopper to ensure no metering unit runs out. Wear gloves for treated seed.

4. Engage agitator if fitted.

5. Fit calibration trays if provided, if not spread large sheet under drill to catch all the grain

6. Fit calibration handle.

N.B. For mounted drills only, raise the drill wheels just clear of the ground on tractor hydraulics.

7. Fill all the metering mechanisms by turning handle about 4 revolutions. Check flow in see-through panel on drills that have this.

8. Empty seed collected in calibration trays or sheet back into hopper.

9. Look at the manufacturer's instruction book to find out the number of turns of the calibration handle required to cover 0.04 ha (or 0.1 acre) (bearing in mind the size of the drive cog fitted).

N.B. Some drills have a variety of gears into which the calibrating handle fits which will affect number turns. Check in manufacturer's instruction book.

10. Turn handle required number of turns.

11. Collect and weigh seed accurately.

12. Multuply by 25 to give weight per hectare (or by 10 for weight per acre).

N.B. Some drills use only part of their width for calibration purposes. In this case divide weight of seed collected by the number of rows calibrated and multiply result by number of rows across full width of the drill before calculating total weight drilled.

13. If wrong, reset drill and repeat procedure.

14. Disengage drive to seed hopper or empty hopper metering mechanism, tubes and coulters.

15. Repeat precedure for fertiliser.

16. Record settings to make calibration easier in future.

It is better to set the drill under the required rate for the first calibration in that it is easy to open up the metering units when full but to try to close them may cause misalignment of the feed mechanisms.

If no calibration handle is provided the same results can be achieved by rotating the land drive wheel as follows:

1. Securely jack land drive wheel clear of ground carefully and safely and chock the other wheel.

2. Calculate the number of turns of the land drive wheel as follows:

Number of turns for 0.05 ha =

$$\frac{0.05 \times 10,000}{\text{drilling width} \times \text{circumference of the wheel (metres)}}$$

where the drilling width = number of rows x row width

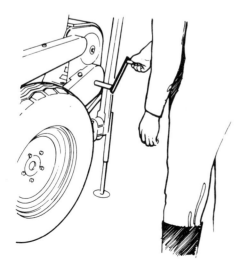

OR

Number of turns for 0.05 acre =

$$\frac{0.05 \times 4840}{\text{drilling width} \times \text{circumference of the wheel (yards)}}$$

The effective circumference of the wheel can be found by:

— marking the land drive wheel at its bottom centre with chalk and making a chalk mark in the same place on the floor

— driving drill forward one wheel revolution until chalk mark is in the same position

— marking floor in line with mark on wheel

— measuring distance between two chalk marks on the floor.

3. Turn the wheel 4 times to fill the metering mechanisms.

4. Collect grain and tip back into hopper. Zero area meter.

5. Turn wheel correct number of turns.

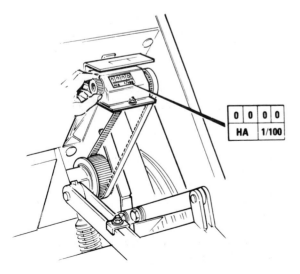

6. Collect and accurately weigh the grain.

7. Multiply by 20 to give rate in kg/ha or lb/acre.

ALWAYS RECALIBRATE IF SEED OR DRESSING CHANGES

b) Field calibration

The application rate in the field may vary after stationary calibration due to:

— wheel slip

— bouncing

— changes in flow characteristics of sample.

To check this, carry out the calibration as follows:

1. Look in manufacturer's instruction book to find distance that is travelled by your width of drill to cover 0.4 ha, or 0.1 (1/10) acre.

   If no distance recommendations are given then work out distance to drive as below:

   i) Distance travelled in metres to cover

   $$0.4 \, \text{ha} = \frac{400}{\text{drill width}}$$

   Where drill width = number of rows x row spacing (metres)

   ii) Distance travelled in yards to cover 0.1 acre

   $$= \frac{484}{\text{drill width (yards)}}$$

2. Measure out this distance.

   Pacing is not accurate enough, so use a measuring chain or alternatively a simple line can be made by:

   — hammering 2 x 150 mm wire nails into a length of wood 1 metre apart, leaving about 100 mm of the nail sticking out

   — tying one end of a length of baler twine around one nail

   — winding the twine around the nails until the right length has been wound out.

   —fixing one end of the twine around a clearly seen peg at one end of the field

   — running the line out up the field and marking clearly the other end of the run.

3. Fill the hopper with sufficient seed.

4. Ensure calibration trays are in place.

5. Fill mechanism as before.

6. Zero the areameter.

7. Drive the required distance, dropping the coulters into work about 1 m before the first peg and lifting them as you pass the peg at the other end.

8. Collect material and weigh it accurately.

9. Multiply by 25 to give rate in kg/ha. (or x 10 for lb/acre).

10. Compare with stationary check, making any alterations in settings necessary and recording discrepancies.

11. Repeat for fertiliser.

12. Wind up the line and keep it with drill for future calibrations.

13. Preparing drill for transport

Implements above a certain width cannot be transported on roads (see 'A Guide to the Use of Farm Vehicles on the Road' M.1.A.5TA) and must therefore be towed lengthways either by transport wheels and a tow bar or on a transporter.

Do not transport with grain or fertiliser in the hoppers, as this adds weight, and also compacts in the hoppers and metering mechanisms.

a) Transport wheels and towbar

Check condition of transport wheels, frame and towbar.

Ensure tyres are in good condition and at correct pressures.

Do not use any pins other than those provided with the machine.

Ensure drill is safely lifted and supported before attaching wheels

△ Do not attempt to man-handle the drill.

△ Ensure all pins are inserted correctly and secured with lynch pins.

△ Stow the field drawbar in a safe and secure position.

△ Stow all hydraulic couplings and electric cables safely and securely.

b) Transporter

Check condition of transporter, especially the tyres.

Position transporter on firm level ground if possible.

Block transporter wheels to prevent them moving during loading.

Load drill onto transporter.

Secure drill effectively.

Unhitch tractor from drill and hitch onto transporter.

Stow the drill drawbar in safe position.

△ Be aware of the drill overhead clearance, particularly with long markers in the vertical position.

△ Drive carefully at all times.

△ Select low gear on steep downhill slopes.

△ Slow down on rough ground.

△ Avoid sudden braking.

△ Never allow passengers to ride on the tractor or drill during transport.

You should understand

— factors affecting choice of tractor to match drill
— the importance of having the drill at correct operating angle
— importance of having coulters in good condition and springs set at correct pressure
— importance of using accurately set bout markers and how to work out their settings
— reasons for tramlining and system employed on your farm
— reasons why drill settings must be changed for different sizes of seed
— importance of accurate setting and calibration of drill
— significance of ensuring that drill is correctly prepared for transport.

If you do not understand all these points, ask your employer, manager, instructor, or course tutor for advice.

Further experience

You will become proficient only by practice. Try to get experience as soon as you can by taking every opportunity offered by your manager or employer to follow up the training you have received.

Proficiency tests

Following initial instruction and further experience you will become eligible to take National Proficiency Tests. The proficiency test to which this guide relates is MO4 Sowing/Planting. You should, however check the test schedule and ensure that you are experienced in the other activities included in the test.

Remember that the board and your county agricultural college offer a wide range of courses in practical skills, together with associated knowledge, which will help you understand more about the work you are doing.

26 POST SEASON MAINTENANCE OF SEED DRILLS

Your aim

is to be able to:

— select and wear appropriate protective clothing when working on drills that have been in contact with seed dressing
— handle and store toxic chemicals safely and dispose of their packaging
— empty and clean drills
— label and store unused materials
— dismantle parts of drill and store correctly
— reassemble parts removed for cleaning
— inspect drill and replace worn or damaged parts and ensure that spare parts are ordered in time for fitting
— lubricate and protect mechanisms as recommended by the manufacturer
— put the drill and dismantled parts into store as recommended
— record spares fitted and those required
— carry out all work relating to safe storage, minimising risk to yourself, other people, livestock and the environment.

You will need

— your farm's drill
— a tractor of suitable power
— manufacturer's instruction book and parts list
— set of hand tools, including those supplied with drill
— tyre pressure gauge and footpump, or airline incorporating a tyre pressure gauge
— grease gun, oil can, lubricants
— cleaning materials, brushes, rags, soda, detergents, dewatering fluid and diesel oil
— preservatives, paint and brushes, rust preventative and gauntlets
— hydraulic jack
— axle stands
— spares of commonly wearing parts
— calibration trays, tarpaulin or sheet
— protective clothing as required by the Health and Safety (Agriculture) (Poisonous Substances) Regulations, 1975
— first aid equipment which complies with the Health and Safety (First Aid) Regulations, 1981
— dust sheet
— Post season maintenance checklist (M.7.A.5CL)
— waste ground away from water supply, ditches etc.

You may need

— length of wood (to place under coulters before lowering).

You should

have received training or gained experience in:

— the safe handling of chemicals
— the use of rust preventative materials
— the preparation of equipment for storage.

Note: △ = important safety point.

Introduction

Use checklist M.7.A.5CL to record maintenance carried out, further work necessary and spares required.

1. Emptying the drill

△ Ensure that you are wearing the prescribed protective clothing if the drill has been used with toxic seed dressing, pesticide or other chemicals.

Fit calibration trays or bags, or position over a sheet in a dry place.

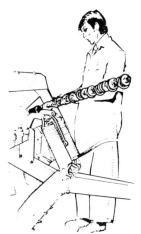

△ Ensure adequate ventilation when emptying the seed hopper.

Remove feed roller or lower feed gates if necessary (see manufacturer's instruction book).

Open cover plates in bottom of hopper if fitted. Clear all seed from hoppers, feed mechanisms and coulter tubes.

Empty material from trays or sheet into original container. Seal well and ensure that it is labelled correctly.

Store seed in a safe vermin-proof place.

Repeat for fertilizer hopper.

2. Washing the drill

Tow the drill to waste ground where unauthorised people and animals cannot get access.

△ Ensure that any washing water cannot drain or seep into water supplies, rivers, streams or ditches.

Hose machine down with cold water (unless manufacturer's instruction book specifies otherwise), removing deposits of fertiliser and seed dressing which might otherwise harden and cause damage.

Do not direct pressure hoses onto bearings.

Pack bearings with grease before pressure washing.

Consult manufacturer's instruction book before using a steam cleaner on any drill.

Run the fan on pneumatic drills to aid cleaning and drying.

△ Do not wear loose fitting clothes when working with p.t.o. driven machines.

△ Keep clear of any moving parts during washing. Dismantle moving parts where appropriate (see manufacturer's instruction book).

Scrub metal parts with hot soda water (250g washing soda in 5 litre water).

Scrub rubber and plastic parts (e.g. coulter tubes) and ensure that there are no deposits of fertiliser or seed dressing left on them.

N.B. Some plastics will distort in hot water. Consult manufacturer's instruction book.

Rinse in cold water.

Allow machine to dry overnight or apply dewatering fluid. Wear gloves if necessary.

3. Checking drill parts

Check drill over thoroughly, carry out necessary maintenance where possible.

Record maintenance carried out and list spare parts removed for store.

List further repairs required and ensure that spare parts are ordered. (Know the procedure on your farm for ordering spare parts). Inform your manager of all work carried out and what further work is required.

Check hopper and the dividing plate for damage, holes or any trapped materials.

Inspect metering mechanism and check for wear of:

— any chains and sprockets

— rollers, peg or star wheels.

Check:

— free movement of controls

— correct adjustment of clutches.

Inspect coulter assemblies:

(a) Disc type:

⚠ — check for play on bearings by moving from side to side. Always wear strong gloves as discs can become very sharp

— check disc rivets and tighten or replace any that are loose

— measure diameter of disc to ensure that it is not worn beyond the manufacturer's limit.

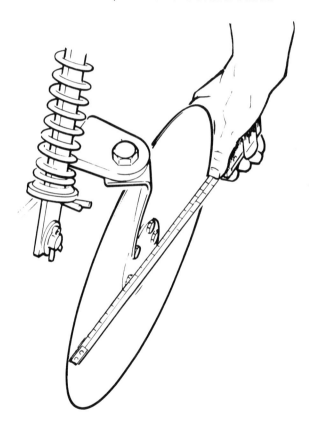

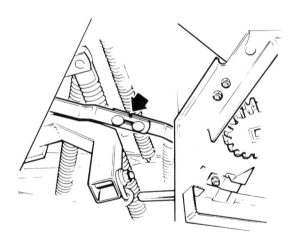

Inspect feed tubes for splits or signs of wear.

Check that feed flaps on bottom of metering units are free.

— rotate discs to check free movement

— inspect condition of scraper or shoe and replace as necessary

— adjust so that it almost touches disc without stopping its free movement.

(b) Suffolk type coulters

Check the leading edge of coulter for wear and replace as necessary.

(c) Hoe or tine type coulters

Check points for wear and reverse them or replace if unserviceable.

Check coulter pressure springs for signs of damage. Release tension before storing.

Ensure that telescopic coulter tubes can telescope freely.

Check that depth control mechanisms work properly and that all adjustments are satisfactory.

Check conditions of the points on the track eradicators and turn or replace as necessary.

Check for wear and damage to any cultivation equipment and ensure free movement of all parts.

Inspect marker tines or discs for wear or damage.

Ensure correct functioning of all controls and moving parts.

Check covering mechanism for damage or wear.

Check insecticide, herbicide or fungicide applicator, where used (see manufacturer's instruction book).

Disconnect all hydraulic couplings. If you are unable to remove a coupling from the spool valve, stop the tractor engine and operate the hydraulic lever to remove pressure; the coupling should now release.

Inspect all hydraulic pipes and couplings for damage. Replace immediately any that show signs of deterioration.

Clean the couplings and secure to the plate on the drill (if fitted). If not, place a polythene bag over the couplings and secure it to prevent dirt getting in.

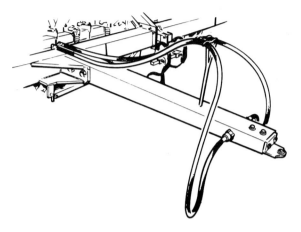

Inspect all electrical cables and bind any damaged insulation with insulating tape. Ensure that cables are re-routed to prevent similar damage happening again.

Check ram seals for leaks and arrange with your manager for the necessary repairs to be carried out.

Inflate tyres to the correct pressures and ensure that dust caps are fitted.

Check that wheel nuts are tight; some require a specific torque (see manufacturer's instruction book).

Check condition of tyres.

Check nuts and bolts on whole machine and see they are tight (check manufacturer's instruction book for correct torque settings).

△ Check that all guards are securely in place.

Check that breakaway device on bout markers works correctly and adjust the spring tension if necessary.

4. Pneumatic drills and those combined with rotavators

In addition to the above check:

(a) P.t.o. coupling

Check splines for wear or damage.

Check condition of shaft and test joints for wear.

Inspect guards to ensure there is no damage and that they telescope freely.

Secure shaft safely.

(b) Fan

Check blades for wear or damage.

Ensure that inlet and outlet ducts are unobstructed.

Ensure free movement.

Inspect condition of bearings and check for play.

Check that the drive to the fan is in good condition and properly adjusted. (See manufacturer's instruction book.)

Release tension on fan belts and remove for storage.

5. Lubrication and rust prevention

Follow manufacturer's recommendations for lubrication.

Ensure you know where all lubrication points are and the correct lubricant to apply.

Grease all nipples by:
— wiping outside of nipple first
— applying two or three pumps only or sufficient just to flush the bearing
— wipe away excess grease from outside of bearing
— rotate mechanism to coat all surfaces of the bearing with clean grease.

Replace any damaged or blocked nipples.

△ Clean all chains and sprockets using diesel oil or paraffin, always wear gloves.

△ Be aware of fire risk; do not smoke and ensure adequate ventilation.

Oil chains and sprockets sparingly; excessive oil only attracts dust, to form an abrasive mixture.

Check gearbox oil level and top up if necessary.

Change the oil in the gearbox if recommended by the manufacturer.

Clean off dirt, oil or rust from any unprotected metal surface, using diesel oil or paraffin and wipe dry.

Paint over any body work where paint has been damaged, using a rust preventative paint.

⚠ Apply rust preventative to any non-painted metal parts. Always wear gloves or protective clothing as specified by the manufacturer.

Take care not to get rust preventative on any plastic, nylon or rubber parts, as this may cause deterioration.

6. Storing rubber, plastic or nylon parts

Wash in a suitable detergent to remove oil and dirt.

Rinse in cold water and dry.

Store away from sunlight or oil contamination, both of which cause deterioration.

Ensure that parts are labelled so you can find them when you next need them.

Ensure that rubber or nylon parts do not come into contact with oil, diesel fuel, neat detergents or rust preventatives.

7. Storing dismantled metal parts

Store in an oil bath or coat with a rust preventative. Store feed rollers in the hopper.

8. Storing the drill

Store the machine under cover in a weatherproof building.

Cover with a sheet to prevent contamination from dust and bird droppings.

⚠ Jack the machine up, place axle stands under secure points and lower carefully onto the supports to minimise tyre deterioration.

⚠ Ensure that machine is safe on its supports.

Relax tension on springs, chains and belts, unless manufacturer specifies otherwise.

Ensure that coulter tubes are not kinked.

If coulters are stored in lowered position, place a board under them and lower them onto this.

Prop hopper lids open fractionally to allow ventilation but not enough to allow vermin in; they can seriously damage plastic parts.

If the drawbar obstructs a passage or is a potential hazard, secure it in its upright position.

Ensure time is allocated to carry out any further maintenance or repair work not completed.

If the drill must be stored outdoors, it must be kept dry. Cover with waterproof tarpaulin and secure it well.

Remove feed roller and store under cover.

You should understand

△ — regulations relating to handling of chemicals on the farm and protective clothing requirements

— importance of removing all materials from the drill at the end of the season

— potential damage that vermin can cause

△ — importance of secure storage of chemically treated seed

— importance of keeping rubber and nylon parts free from oil, diesel or paraffin

— importance of a systematic maintenance procedure at the end of season

— why it is essential that the drill and its parts are stored correctly.

If you do not understand all these points, ask your employer, manager, instructor or course tutor for advice.

Further experience

You will become proficient only by practice. Try to get experience as soon as you can by carrying out the end of season maintenance to your own drill at the end of the next season.

Proficiency tests

Following initial instruction and further experience you will become eligible to take National Proficiency Tests. The proficiency test to which this guide relates is MO4 Sowing/Planting. You should, however check the test schedule and ensure that you are experienced in the other activities included in the test.

Remember that the Board and your county agricultural college offer a wide range of courses in practical skills, together with associated knowledge, which will help you understand more about the work you are doing.

27 PREPARING FIELD CROP SPRAYERS

Your aim

is to be able to:

- attach a mounted or trailed sprayer correctly to a tractor

- clean the sprayer to ensure that all traces of chemical are removed

- check and test the sprayer for correct operation, replacing faulty hoses, clips and nozzles as necessary

- calibrate a sprayer so that it delivers the required quantity of spray per hectare.

You will need

- a mounted or trailed field crop sprayer in good working condition

- a tractor of adequate power and weight to operate it safely

- instruction books for tractor and sprayer

- 'Instructions for Use' for the spray chemical that you will be using

- spare check-valves and new nozzles of differing sizes

- hosepipe and long-handled brush

- measuring jug and jar

- notepad and pencil

- copy of booklet 'The Safe Use of Poisonous Chemicals on the Farm'

- copy of 'Crop sprayer calibration' training aid (M.9.D.1.TA)

- copy of booklet 'Code of Practice for the Disposal of Unwanted Pesticides and Containers on Farms and Holdings'

- all necessary protective clothing as required by 'Instructions for Use' for the chemical used, or in the leaflet 'The Safe Use of Poisonous Chemicals on the Farm'

- washing soda/detergent according to chemical used

- weigh scale

- tape measure

- first aid equipment which complies with the Agriculture (First Aid) Regulations 1957.

You should

have received training or gained experience in:

- attaching p.t.o. shafts (trainee guide M.1.B.3)

- manoeuvring with mounted machines (trainee guide M.1.B.5) or

- manoeuvring with trailed machines (trainee guide M.1.B.6).

Note: △ = important safety point

1. Introduction

(a) *When to prepare your sprayer*

A sprayer which has not been used for longer than two months will need thorough checking before use:
- prepare the sprayer in advance of use
- allow time for replacement spares to arrive.

(b) *Safety*

Even though the sprayer may not have been used for many months, poisonous chemicals may still be present. When washing the sprayer:

△ — never put a nozzle or any other part to your mouth

△ — never allow washing water to run into sewers, ditches or domestic water supplies.
Refer to Code of Practice for Disposal of Unwanted Pesticides and Containers for full information

△ — wear rubber gloves and the necessary protective clothing as specified in the booklet 'Safe Use of Poisonous Chemicals on the Farm'.

2. Attaching to tractor

(a) *Mounted sprayers*

△ Check with sprayer and tractor instruction books that the tractor is heavy enough to carry the sprayer safely:
- fit front end weights if necessary.

Before fitting a p.t.o.-mounted pump, check:
- pump shaft rotates freely

Photographs by courtesy of British Farmer and Stockbreeder

- oil level is correct (on a diaphragm pump)
- rollers on a roller vane pump are free and in good condition
- air pressure in the pressure accumulator is correct (on some diaphragm pumps).

Attach the p.t.o. mounted pump before hitching the sprayer to the tractor:

- attach pump stabiliser chain to a fixed point on the tractor, e.g. drawbar; not to the lift arms or top link
- ensure that pump and p.t.o. guards are correctly fitted.

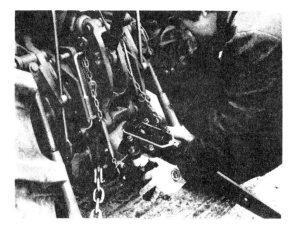

Attach the sprayer to the tractor and raise it to the correct height:
- follow the sprayer manufacturer's instructions, especially where an unusual top link is used
- fit stabilisers to prevent side-sway
- connect the p.t.o. shaft to the tractor (where the pump is mounted on the sprayer).

Check that both tractor rear tyres are at the correct pressure, then ensure that the sprayer is level from front to rear and from side to side:

— measure the tractor lift rods and adjust the right hand one until both are equal.

(b) *Trailed sprayers*
Connect drawbar and p.t.o. shaft following the same procedure as for other trailed machines. Adjust the controls so that they can be reached from the tractor seat, and ensure that they do not touch the cab at any time.

If the sprayer is fitted with a remote control block for use in Q-cabs, fit this in the recommended way and check that all cables and pipes run freely to the sprayer, and do not touch the drawbar or p.t.o.:

— check remote controls, when installed, for correct operation.

3. Cleaning the sprayer before use

The sprayer should always be washed out before use, even though it may have been cleaned before storing.

Even a trace of an unsuitable chemical can cause damage to the crop.

Nozzles, check-valves, filters and boom end caps should have been removed and stored in the tank filter basket:

— if they are still fitted, remove them and place in a bucket of water with a little detergent.

Wash out the tank, taking care to clean inside the tank roof, using a brush:

— remove the drain plug and flush the tank out

— tank washings should run into a soil soakaway (see 'Code of Practice for the Disposal of Unwanted Pesticides')

Refill the tank and pump out through the booms (nozzles removed).

Wash the nozzles, check-valves and filters by gently using a brush (not a wire brush) and swishing in the water, or compressed air:
— inspect the sealing rings. Replace if cracked or stiff or flattened
△ — do not blow through nozzles
— inspect the check-valves to ensure that they move and seat freely and that there is no dirt on the valve seats.

If tank sight tube is badly discoloured, replace it.

4. Testing the sprayer

Refit the filters (replace with new if they are damaged or cannot be changed.

Replace the check-valves and nozzles:
— it is good policy to fit new nozzles at the start of each season if large areas are to be sprayed
— ensure that nozzles are the correct size for intended application rates and speed of travel
— fit fan nozzles with the slot in line with the boom.

Fill the tank and spray out. Inspect the sprayer whilst it is spraying:
— check all rubber pipes by flexing them while under pressure. Any leaking pipes must be replaced, and all hose clips must be fully tight.

Look inside the tank. A fine froth on the surface shows that the pump is sucking air. Search carefully from the tank to the pump to find the air leak.

Vary the pressure regulator settings and check that the gauge reading changes and also that it reads zero when the pump is stopped.

Change from 'spray' to 'spray off' and back several times to ensure that the controls work properly:
— make sure all check-valves work. Any dripping nozzles should be inspected and the valves cleaned or changed

Inspect the nozzle patterns whilst spraying:
— hold a large piece of black card behind each nozzle to help you to see the spray patterns.

Replace any nozzle which shows streaks in the spray.

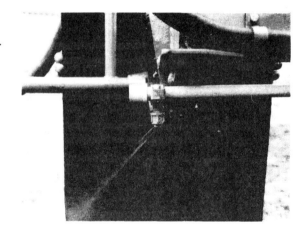

Check the nozzle outputs:
— all nozzles must be spraying properly before starting the checks

— hold a jar under each nozzle in turn whilst spraying. Write down the time taken for each nozzle to fill to a mark on the jar.

— from the list of times, find the average figure. Scrap those nozzles which are more than 5 per cent from the average.
e.g. average time 40 seconds. 5 per cent of 40 seconds = 2 seconds. Therefore replace those nozzles whose times were less than 38 seconds or more than 42 seconds.

Variations may be due to wear, partial blockage, damage or wrong size.

Refer to manufacturer's instructions for the correct height from nozzle to target. The target may be the ground, the top of the crop or the top of the weeds:
— adjust the boom on its mounting frame or alter the height of the entire sprayer on the tractor
— check also the height of the outer boom sections

Test for striping
— spray along dry concrete or tarmac
— spray from every other nozzle should just overlap at the target

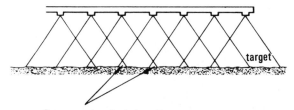

Spray overlaps just above target.

— stripes when drying show that the nozzles are too low
— even drying shows the nozzle height is correct for ground spraying. When spraying crop or weeds, nozzles should be the same height above the target in the field.

5. Calibrating the sprayer

Calibration should always be carried out:

— at the beginning of each season

— after every 100 hectares (250 acres)

— after changes of tractor or wheels, nozzle tips or operating pressure.

(a) *Sprayers without regulating discs* — see training aid 'Crop Sprayer Calibration' (M.9.D.1.TA)

When accurately calibrated, write on a tank label:

— which tractor is used

— tractor gear and engine revs

— spraying pressure

— nozzle size

— application rate.

(b) *Sprayers using regulating discs*

Your aim is to collect the output from one nozzle over 100 metres and use this and the manufacturer's chart to find out which discs should be used:

— accurately measure 100 m in a field

— fit the two calibrating discs specified in manufacturer's instructions. Check that all nozzles are spraying correctly, then test spray the measured 100 metres.

Have an assistant walk behind the boom, collecting the output from one nozzle over 100 m in a jar.

By referring to the manufacturer's chart, the right discs for the required rate can now be fitted.

These particular discs are only correct for that tractor, in the one gear, and with the same nozzles fitted. Changing any of these will mean re-calibrating.

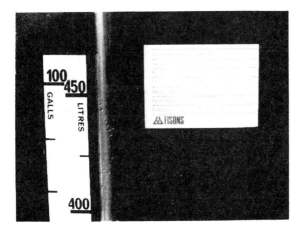

You should understand

△ — why thorough washing is essential before using a sprayer

— that the cost of new nozzles is small compared with the cost of chemicals used

— the effects on the crop of faults in spraying

— the importance of recording details of settings during calibration

△ — relevant agricultural safety legislation, including Health and Safety (Agriculture) (Poisonous Substances) Regulations 1975, Health and Safety at Work etc. Act, 1974, Agriculture (Field Machinery) Regulations, 1962, Agriculture (Power take-off) Regulations, 1957.

Further experience

You will become proficient only by practice. Try to get experience as soon as you can by taking every opportunity offered by your manager or employer to follow up the training you have received.

Remember that the Board and your county agricultural college offer a wide range of courses in practical skills, together with associated knowledge, which will help you understand more about the work you are doing.

28 CALIBRATING FIELD CROP SPRAYERS

Calibration

Calibration should be carried out:
— *at the beginning of each session*
— *after every 100 hectares (250 acres)*
— *after changes of tractor or wheels, nozzle tips or operating pressure.*

1. Check chemical packs for any special instructions, then choose an application rate within those recommended, e.g. 200 litres per hectare (20 gallons per acre).

2. Choose and fit nozzle tips to operate in their acceptable pressure range appropriate for the chosen application rate and anticipated forward speed.

3. Carry out a trial run with the sprayer tank half full of water spraying on a surface similar to the average on which you will spray. Check that the proposed forward speed gives an acceptable level of boom bounce and boom yaw (back and fore movements) and the gear selection gives a pto rev/min of about 540.

4. On the same surface check tractor speed:
   (a) In midfield mark out a distance of 100 m(yds) with two canes; pacing is not accurate enough — use a tape.
   (b) Select the gear and engine rev/min as in (3).
   (c) Measure the time taken to travel 100m(yds) spraying at the selected engine rev/min. Start and stop timing as tractor knocks canes over.
   (d) Refer to the following tables to check the approximate forward speed in kilometres per hour (miles per hour).

Time in Seconds	120	90	72	60	51	45	39	36	33	30
Km/h	3	4	5	6	7	8	9	10	11	12

Time in Seconds	102	82	68	58	51	45	41	34	29	26
mph	2	$2\frac{1}{2}$	3	$3\frac{1}{2}$	4	$4\frac{1}{2}$	5	6	7	8

   (e) Record tractor registration number, tyres fitted, gear, engine rev/min time for 100m(yds) and aproximate forward speed.

5. Return to yard. Adjust spraying pressure to the level recommended in the chemical instructions and/or within the nozzle chart range.

6. Check nozzle spray patterns and alignment visually, replace any rogue nozzle with a nozzle matched to the flow rate of others in the group.

7. Compare individual nozzle outputs by nozzle flow measure or recording the time required to fill a measure to a pre-determined level. Replace any nozzies more than ±5 per cent from the average.

8. Turn 'Spray On' to clear air from the boom, then 'Switch Off' pump. Ensure the tank is level and fill it with clean water to halfway up the neck of the tank — mark the level inside the neck front and rear.

9. With the sprayer stationary and the tractor rev/min at the setting used when the forward speed was checked, spray out for the time taken to travel 100m(yds) in 4(c).

10. Refill the sprayer to the marks in the neck with clean water, using a calibrated measuring vessel or flow meter.

11. Establish the effective swath width of the sprayer in metres i.e. the distance between a pair of nozzles x the number of nozzles.

12. Compare the quantity of water required with that shown on the calibration chart or the chosen application rate and the effective spraying width of the sprayer.

13. If the rate is not correct:
    — make small adjustments by varying pressure. NB Do not go outside chemical and/or nozzle tip recommendations.
    — make large adjustments by changing nozzle tips.

14. Re-check after each adjustment until correct.

15. Record for future use:
    — nozzle tips fitted
    — application rate
    — spray pressure

    together with information from 4(e), and date of check.

Calibration Chart – METRIC

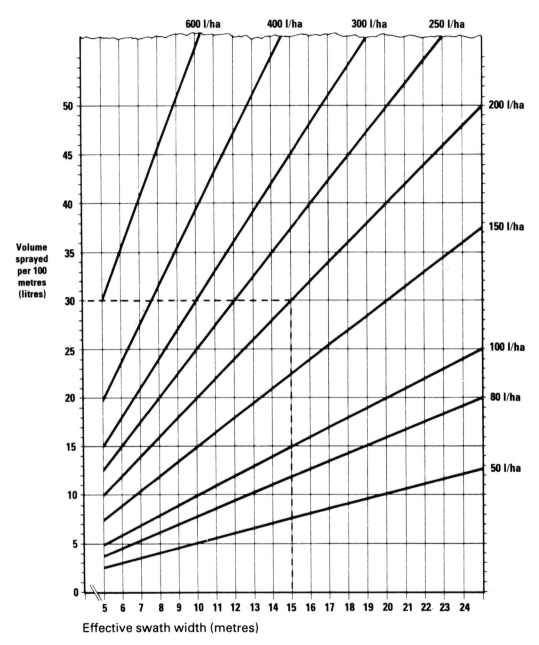

Example (see dotted line):- A sprayer with a swath width of 15m at 200 l/ha should deliver 30 litres over 100m test-run.

Calibration Chart – IMPERIAL

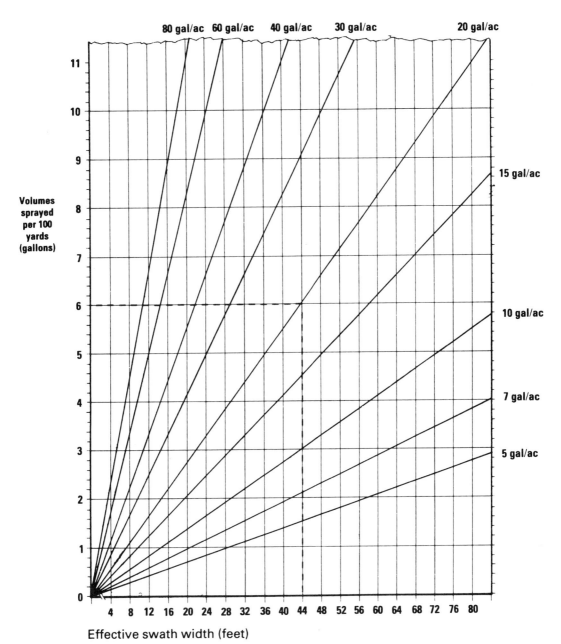

Example (see dotted line):- **A sprayer with a swath width of 44ft at 20 gal/ac should deliver 6 gallons over 100 yd test run.**

Equipment check list

Tractor

Sprayer (having undergone pre-season preparation)

Tractor instruction book

Sprayer instruction book

Sets of nozzle tips, spares and instructions

Small measuring vessel or nozzle flow meter

100m(yds) tape and two canes

Large calibrated measuring vessel or hose flow meter

Calibration chart

Watch (with sweep second hand) or stop watch

Supply of clean water

A hard level surface

A typical field with a 100m(yds) run.

This calibration procedure was agreed and adopted by the AEA Chemical Application Machinery Manufacturers Group in September, 1979 and was first demonstrated at Fisons Sprayers '79 by the ATB.

29 CROP SPRAYING

Your aim

is to be able to:

— fill the sprayer tank with water and chemical safely and in the right proportions, and ensure adequate mixing

— dispose of empty containers and unwanted pesticides in accordance with M.A.F.F. leaflet 'Code of Practice for Disposal of Unwanted Pesticides and Containers on Farms and Holdings'

— determine whether the chemical being used is a 'specified substance' under Health and Safety (Agriculture) (Poisonous Substances) Regulations 1975, and if so to comply with all the requirements of the regulations in respect of protective clothing, training, hours of work and keeping a register and notification of sickness

— in the case of non-specified substances, follow the manufacturers' instructions regarding safety and protective clothing

— operate a sprayer, using markers, without overlap or missed strips, at the right pressure, ensuring that all nozzles are working correctly, so that the required application rate is achieved over the whole field

— wash out the sprayer after work thoroughly, so that it will be ready for further work the following day

— de-contaminate and apply preservative to the sprayer before prolonged storage.

You will need

— a mounted or trailed field crop sprayer, in good working condition, on which you have carried out full preparation and calibration as in trainee guide M.9.D.1, attached to the tractor which was used for calibration testing

— manufacturers' instructions for tractor and sprayer

— 'Instructions for Use' for the chemicals you will be using

— spare check-valves and new nozzles of differing sizes

— hosepipe and long-handled brush

— notepad and pencil

— copy of booklet 'The Safe Use of Poisonous Chemicals on the Farm'

— copy of booklet 'Code of Practice for the Disposal of Unwanted Pesticides and Containers on Farms and Holdings'

— all necessary protective clothing as required in 'Instructions for Use' for the chemical used, or in the booklet 'The Safe Use of Poisonous Chemicals on the Farm'

— an area for washing the tractor and sprayer after use, equipped with hosepipe and having a safe drain which does not run to a ditch, pond or stream

— marker poles (if sprayer is not fitted with foam blob marker)

— first aid equipment which complies with the Agriculture (First Aid) Regulations, 1957.

You must have

received training or gained experience in:

— preparing field crop sprayers (trainee guide M.9.D.1).

— calibration (training aid M.9.D.1.T.A).

Note: △ = *important safety point.*

1. Introduction

△ Always read and follow the 'Instructions for Use' provided with all spray chemicals.

△ Do not use chemical from an un-labelled container.

△ You must wear the protective clothing specified in the 'Instructions for Use'.

△ If a specified chemical is used, you must wear the protective clothing as required by the booklet 'The Safe Use of Poisonous Chemicals on the Farm'.

2. Mixing and filling the tank

△ Always read the label before opening a chemical container.

△ Dress in protective clothing as specified in 'Instructions for Use' or 'The Safe Use of Poisonous Chemicals on the Farm' before opening a chemical container.

△ Check that weather and crop conditions are suitable for spraying.

Work out how many hectares a tankfull of spray will cover by dividing tank capacity by application rate:

e.g. $\frac{500 \text{ litre tank}}{200 \text{ litres/hectare}}$ = 2.5 hectares per tankfull.

or $\frac{100 \text{ gallon tank}}{20 \text{ gallons/acre}}$ = 5 acres per tankfull.

To work out how much chemical to add to a tankfull of water:

— from 'Instructions for Use' find the dose rate of chemical per hectare
— multiply this by the hectares covered per tankfull

e.g. dose rate — 7 litres per hectare
at 2.5 hectares per tankfull
chemical per tankfull = 7 x 2.5 = 17.5 litres of chemical.

or dose rate — 5 pints per acre
at 5 acres per tankfull.
chemical per tankfull = 5 x 5 = 25 pints of chemical.

Start filling the tank with clean water:
— if using self-fill attachment, ensure that no sand is allowed to enter the suction pipe.

— do not run a roller vane pump dry. Wait until the tank has started to fill, then start agitation

— check agitation inside the tank whilst filling, before adding chemical

— a mobile tank in the field with a large filling pipe reduces the time spent filling

— always filter the water through filter basket in the lid

△ — do not make a direct connection between the tank and mains water supply (risk of mains water contamination)

△ — do not fill from cattle drinking troughs

— always remove suction pipe from water source as soon as self-filling is completed.

If using wettable powders:

— follow the 'Instructions for Use' on mixing exactly

— when pre-mixing is specified, weigh out powder into a clean bucket and mix with water until a smooth cream with no lumps is formed

- use the chemical probe to suck the cream or pour it into the tank lid filter, wash the bucket and stirring stick over the filter, and wash all chemicals through the filter with the hose.

If using liquid chemicals:
- measure out the correct quantity into a clean container marked in litres, if you have to split a pack
- pay close attention to 'Instructions for Use' regarding the order in which chemicals and water are added to the tank

Fill to the required level:
- do not make up more spray mix than will be needed
△ — always stay with the sprayer while filling, to prevent overflow
- do not leave the sprayer standing for long with mixed chemical. This can lead to breakdown of chemicals and settling out.

When filling is complete, stir with a long stick to ensure full mixing of chemicals which may be lying on the bottom of the tank or on the surface.

- use the chemical probe or pour the chemical through the tank filter and wash the containers into the filter so that all chemicals and water are added to the tank.

Continue agitating the tank contents unless 'Instructions for Use' state otherwise.

Lock away all unused chemicals:
△ — never pour chemical from its original container into an un-labelled container for storage
- store empty containers in a secure place until they are disposed of.

Empty containers should be kept separate from full ones in a good chemical store.

△ — dispose of containers and unwanted chemical as stated in 'Code of Practice for the Disposal of Unwanted Pesticides and Containers on Farms and Holdings'.

3. Planning field work and spraying the field

Normally spray twice round the headland then work up and down parallel with the longest straight side, or follow the drill rows.

At the corners of the headlands, switch off the spray, reverse into the corner to square it, then switch on and continue. Turning round the corners without stopping and switching spray off causes under-dosing on the outside of the turn and over-dosing on the inside.

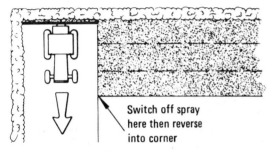

Switch off spray here then reverse into corner

△ Spray drift is dangerous. Take care not to affect nearby bees, crops, stock and wildlife.

△ Do not spray in windy weather when drift will be blown outside the field.

△ Where the tractor is not fitted with an approved air ventilation unit and sprayer remote controls, it may be necessary to open the cab windows to reach the controls and allow ventilation. Cab doors should always be closed.

△ In a light breeze you must spray with the breeze to the side:
 — work with the sprayed area downwind so that spray is not blown onto the tractor

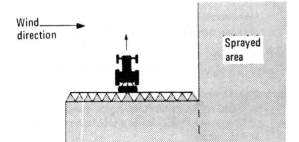

Wind direction → Sprayed area

— this may mean waiting for calmer weather if your direction of spraying is fixed by tramlines or drills.

Mark out the field:
— if markers are placed securely and accurately they can be left in the hedgerow all season
— use tall marker posts, easy to see, e.g. tie a fertilizer bag to each
— place a marker opposite every other run

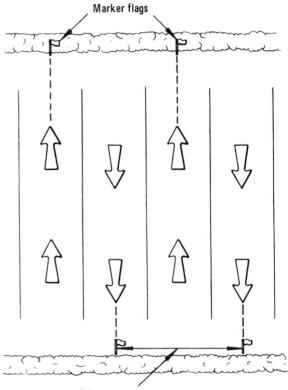

Marker flags

Distance apart = twice swath width

— do not rely on pacing the distances. Use string to measure the distance between markers (twice the swath width) with an assistant to help you.

Tramlines, if accurately placed, mean that you can always match the bouts easily.

If using a foam marker, follow the manufacturer's instructions to ensure that foam blobs are big enough and spaced so that you can follow them easily.

— switch the spray on and off accurately when crossing the headland. Allow for the time required for booms to refill if the sprayer uses a suck-back system. Never spray whilst turning on the headland

Photograph by courtesy of Saltney Engineering Co. Ltd.

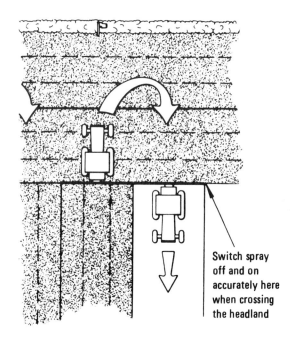

Switch spray off and on accurately here when crossing the headland

Check that nozzle height above target is correct.

Take care to avoid damage to the boom by hitting trees and hedges, especially when cornering.

Check that the pressure gauge reading and engine revs are correct and constant while spraying.

Spray the remainder of the field:

— position the sprayer accurately before starting a bout

— do not run out of spray part way across the field. (If this does happen, stop, reverse round and run back along the wheel marks.)

Keep a regular check on nozzles throughout spraying:

— always carry spare nozzles as replacements

△ — never blow through nozzles to clear blockages

△ — apply tractor handbrake when working on the sprayer.

△ — whenever changing nozzles with a specified substance in use always wear the required protective clothing. If a non-specified substance is being used always wear rubber gloves and face shield

Keeep a constant check on pressure and engine speed during spraying and keep them the same as when calibrating.

Variations in pressure affect:
— application rate
— droplet size (high pressure causes smaller droplets and dangers of drift)
— spray pattern (the angle of the spray from the nozzle)

Recalibration should always be carried out:
— after every 100 hectares (250 acres)
— after changes of tractor, wheels, nozzle tips or operating pressures. See M.9.D.1.T.A.

Use the fault chart to help you to trace the cause of faults which may appear while spraying.

FAULT CHART

FAULT	CAUSE	ACTION
Fine foam on liquid in tank	Air leak between tank and pump, or in pump	Locate and repair leak
Coarse foam on liquid in tank	Too much agitation during filling	Continue filling with pump off
		Extend return flow pipe to bottom of tank
Streaks in spray fans or cones	Nozzle partially blocked	Clean nozzles with compressed air
	Nozzle worn or faulty	Replace nozzle
Narrow spray fans or cones	Pressure too low	Increase pressure setting within recommended range
	Blocked suction filter	Remove and clean filter

FAULT	CAUSE	ACTION
Air spluttering from the nozzles. Pressure gauge flickering	Tank almost empty	Stop spraying. Refill tank
	Air leak between tank and pump, or in pump	Locate and repair leak
No spray appears when switched on	Tank empty	Refill tank
	Pump suction filter blocked	Remove and clean filter
	Tank outlet blocked	Disconnect outlet pipe and clear
	Faulty pump	Fit new or re-conditioned pump
	Air in system	Prime the pump. Run pump long enough to expel air
	Nozzles and check valves assembled incorrectly	Re-assemble correctly, following manufacturer's instructions
Spray stops after short time spraying (pressure falling)	Vacuum in tank	Clear tank vent hole
	Pump suction filter blocking quickly	Remove and clean filter. Clean out tank if repeated trouble
Uneven spray pattern across the boom	Some nozzles blocked	Remove and clean nozzles and filters
	Nozzles not all same size	Check number on each nozzle and change those wrong
	Worn nozzles	Replace with new
Spray stops or gradually goes down (pressure rising)	Nozzle filters or delivery line filter blocked	Remove and clean all nozzle and delivery line filters
Spraydrift	Too windy	Stop spraying. Wait for calmer weather
	Droplet size too small (pressure too high)	Reduce pressure within advised limits.

5. After work

Dispose of any unwanted chemical in accordance with 'Code of Practice for the Disposal of Unwanted Pesticides and Containers on Farms and Holdings'

— never leave chemical in the sprayer overnight.

Refill with clean water and dispose of it in a soil soakaway:

— wash the outside of the sprayer and tractor to remove chemical

— if using the sprayer the next day, leave the tank filled with water.

Note:
Certain chemicals require all hoses and pump to be left open to the air. Follow 'Instructions for Use' carefully:

— if there is any chance of frost, drain the tank and pump and remove nozzles and filters.

Carry out daily maintenance on sprayer and pump as detailed by the sprayer manufacturer.

Storing the sprayer

Carry out de-contamination procedure as above, then:

— refill the tank and add a suitable de-watering fluid

— ensure all filters and nozzles are removed, then pump out through the booms

— remove the tank drain plug. Keep the plug, nozzles and filters in the tank lid filter. Leave the tank lid slightly open to allow the tank to dry

— remove hoses and hang them straight, out of sunlight. Plug the open pipes.

Check the pump and follow manufacturer's recommendations on oil change and storage.

Lubricate the sprayer boom hinges, boom winch and other moveable parts and store the sprayer safely, under cover.

Order spare parts now.

Changing chemicals

When changing to a different type of chemical, the sprayer must be de-contaminated:

— refill the tank again and add the correct cleaning material. For spray containing oils use 50 millilitres of liquid detergent per 100 litres of water. For all other sprays use 100-150 grams of washing soda per 100 litres of water

— agitate the tank, then spray out on a soil soakaway

— repeat the wash

— wash through twice with clean water, then remove nozzles and filters.

Further experience

You will become proficient only by practice. Try to get experience as soon as you can by taking every opportunity offered by your manager or employer to follow up the training you have received.

Remember that the Board and your County agricultural college offer a wide range of courses in practical skills, together with associated knowledge, which will help you understand more about the work you are doing.

30 MAINTAINING RECIPROCATING KNIFE MOWERS

Your aim

is to be able to set and adjust any fully-mounted reciprocating-knife mower, with the aid of the manufacturer's instruction book, so that it will cut with maximum efficiency without undue wear and tear.

You will need

- the mower that is used on your farm (if instruction is on your farm)
- the tractor that normally drives it
- instruction books for mower and tractor
- spare knife, and spare knife sections and rivets
- normal workshop hand tools
- anvil or similar hard flat surface (for rivetting)
- rivet snap
- workshop bench with vice
- clean lubricants of correct type
- first aid kit
- one axle stand or similar support
- flat carborundum stone with handle.

You should have

received instruction or gained experience in:

- manoeuvring with a mounted machine (trainee guide M.1.B.5)
- maintaining machine components.

Note △ = important safety point.

Introduction

Mowers vary in detail of construction, particularly in the main frame and lifting mechanisms, but the cutterbar parts are broadly similar.

You need to know the names of the cutterbar parts.

If there are parts whose names you do not know, or if you do not understand how certain components work, be sure to ask your employer, instructor, or supervisor before you start working on the mower.

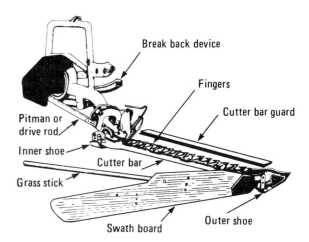

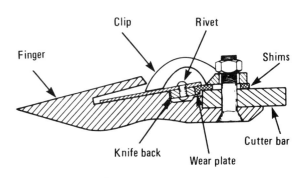

Cross-section of cutter bar

1. Maintaining the knife

(a) *Remove the knife*

It is far easier to remove the knife if the mower is supported on the tractor linkage:

- ensure tractor is parked on firm, level ground
- apply tractor hand brake and stop engine
- △ securely place axle stand or support under mower main frame
- grasp the cutterbar away from the knives and swing the cutterbar into horizontal position
- disconnect the knife drive-link by levering up the locking-clip.

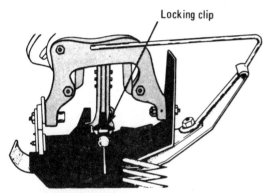

- or unbolting the drive link

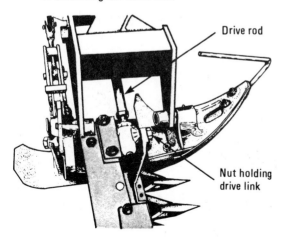

- withdraw knife by pulling on the drive-link connection or,
- if the knife is distorted, an assistant can help by tapping down with a piece of wood, e.g. hammer handle. Secure a length of rope to head of knife to assist pulling knife out
- △ do not allow an assistant to touch knife whilst it is being withdrawn; ensure that he stands clear.

(b) *Check knife back*

Knife back must be straight to avoid being trapped whilst running:

- sight down the length of the knife to check for distortion
- most knives will have a gentle, gradual bow in them
- if this can be pushed flat by hand whilst sighting, the bar is fit for further use
- some straightening can be done by holding the knife-back in a vice and pulling it straight
- if the knife back cannot be pushed flat by hand, it should be scrapped.

(c) *Check for loose knife sections*

Loose sections make a metallic tinkling sound when knife is held vertically and tapped on a concrete floor:

- now check each section to find which are loose
- loose sections must be re-rivetted. The existing rivets will have become weakened by the movement of the loose sections.

(d) *Check for worn or damaged sections*

When sections become worn down to a point they must be replaced. Also replace sections which are:

- worn into a curve on the cutting edges
- chipped badly on the cutting edges
- bent.

(e) *Remove knife sections:*

- hold knife in vice with sections pointing downwards. Vice jaws should lightly grip the sections
- with a heavy blow, shear each rivet of the section. Repeated light taps will damage the section.

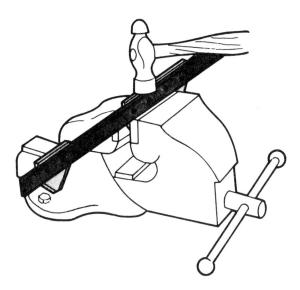

- support the knife back while removing the broken rivets
- carefully examine for damage any sections which you intend to re-rivet
- do not re-use any damaged sections. It is cheaper in the long run to renew them.

(f) *Fit new knife sections*

Ensure that replacements are correct for the machine. Rivets must be a snug fit in section and knife back:

- small rivets will work loose.

Place two rivets in the knife back, fit a section over them and ensure that the section lies flat on the knife back.

Strike the rivet with the face of the hammer to spread the stem of the rivet so that it grips the section.

Use the ball end of a hammer to spread out the rivet:

- beat out the rivet by repeated blows with the ball end of a hammer, working round the rivet as the stem is gradually worked flatter
- finished rivet should be slightly flatter, but just as neat as the head on the other end
- check that newly riveted section is tight, and in line with the other sections.

(g) *Sharpen the knife*

Clamp the knife back in a vice with sections pointing horizontally away from you:

- knife must be held rigidly
- support a long knife at the ends to avoid sagging.

Your aim is to re-create the angles that the sections had when new:

- have a new section for comparison when sharpening
- △ – always sharpen the cutting edge downwards and away from you

Movement of stone when sharpening

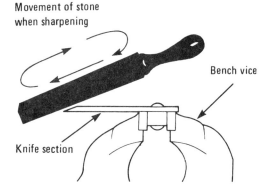

Bench vice

Knife section

△— lift the stone away from knife for the return stroke to avoid cutting fingers

— the cutting angle is correct when sloping surface is the same width as on a new section

— do not tip the stone either steeper or shallower

— ensure that sections are sharpened into the 'V' between sections not just at the tip

— there should always be a flat tip. If the section has worn or been sharpened to a point, replace the section.

2. Checking the fingers

The fingers must be in line at all points where the knife touches:

— tie a thin string around the ledger plates at both ends of the bar, and check that string just contacts all the ledger plates

— if one finger is bent upwards, string will not touch fingers on either side

— to get all ledger plates in line, follow manufacturer's instructions. These may include:

(a) levering fingers with a length of pipe,

or

(b) adding or removing shims from under fingers,

or

(c) replacing with new fingers

— use the string to check if points and tops of fingers are in line

— tops can be gently hammered or levered

— points can be levered gently.

Edges of ledger plates should be sharp. If they are worn, remove fingers and touch-up the sides with a bench grinder or stone.

— new fingers are needed where damage or wear is severe.

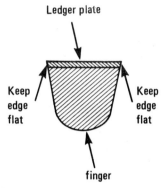

Cross section of finger

Sharpen points if they are worn round.

3. Fitting the knife

Slacken clamp bolts and slide knife clips towards the fingers as far as they will go. Insert knife into cutter bar and slide it fully in:

△— push the knife only by its head. Do not guide the knives, or allow an assistant to guide the knives by pressing by hand

— to allow easier fitting, an assistant, if available, can guide the knives with a piece of wood

— when knife is fitted, check it for free movement.

Knife clips must now be adjusted to hold knife down, and the wear plates must also be adjusted to stop fore and aft movement:

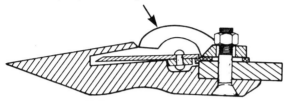

Clip requires bending to touch section

Knife not seated on ledger plate
– add shims under wear plate

— shims under the wear plates raise the rear of knife and tip the front of section down. Add or remove shims until knife sits flat on the ledger plate

— slide clips towards the knife back to press onto sections

— tighten nut when clip stops knife lifting, and wear plate stops knife moving fore and aft

— adjust every clip and wear plate

— when all are adjusted, check that you cannot lift the sections off the ledger plates, and that you cannot push the knife fore and aft

— check that knife will move freely in and out

— if you cannot move it smoothly by pushing and pulling at the drive link end, one or more clips or wear plates are too tight and must be re-set

— after any adjustments always re-check that knife will move freely.

4. Checking knife register

On most mowers at the end of each stroke sections must stop centrally over a finger as in the diagram below:

Knife stops in centre of finger at each end of its stroke

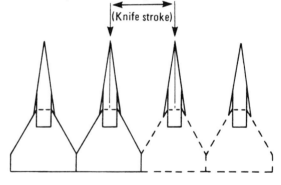

(Knife stroke)

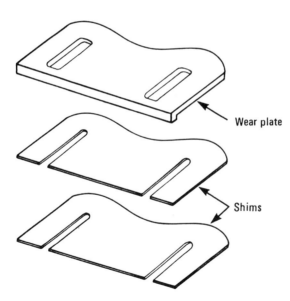

Wear plate

Shims

— at the same time, push the wear plates towards the knife to close the gap behind the knife back

— there should be just enough clearance to stop the knife from binding

— connect the knife drive mechanism, and turn flywheel slowly by hand

— investigate any sticking of the knife and correct as in section 3

— stop the knife at both ends of its stroke and check for correct positioning.

On some mowers, the stroke is not enough to allow knives to centre under the fingers.

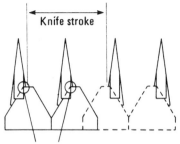

Cutting edge must completely pass over edge of finger

Refer to instruction book for the correct measurement. There should be the same distance from the finger at both ends of the movement:

— refer to instruction book for method of adjusting register. It is usually done by:

— either adjusting length of drive link

— or moving cutter bar in relation to the main frame

Follow manufacturer's instructions.

5. Checking cutterbar lift

When correctly adjusted, the cutterbar should be held with its outer edge several inches higher than its inner edge when fully raised:

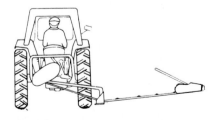

Outer shoe raised higher than inner shoe

— refer to instruction book for details of measurement, and how to correct it, as these vary with different models.

6. Checking cutterbar lead

Cutterbar lead is the distance by which the outer end of the bar 'leads' the inner end when stationary:

— this allows the bar to be dragged back slightly by the grass, yet still be at 90° to the direction of travel

— check by drawing line on floor or with string stretched from tractor tyres across the front of cutterbar

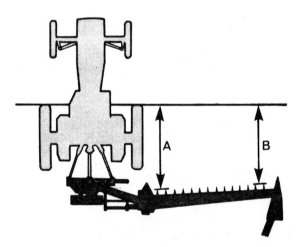

Line A should be longer than line B

— outer edge of bar should be nearer the line

— refer to instruction book for correct setting for your model of mower

— if in doubt, the amount of lead is usually 20mm per metre length of cutter (¼ inch/foot) or 2 in 100 (2 per cent)

— methods of altering lead vary, so follow manufacturer's instructions.

7. Checking break-back

All mowers are designed to allow the cutter-bar to swing back when an obstruction is hit:

- break-back is spring loaded so it should hold cutterbar when in normal use
- test break-back by pulling the outer end of the cutterbar with a rope. It should just be possible to release the cutterbar
- lubricate the mechanism and re-set it
- if in doubt, set break-back to release too easily, and gradually tighten up during work in the field so that it just holds in work.

8. Checking belts and lubricating

Always lubricate the areas where you are working (if applicable) rather than go round the entire machine after:

- follow manufacturer's instructions on frequency of lubricating and types of oil and grease required
- check drive belt tension and pulley alignment as specified by manufacturer
- do not let split oil or grease remain on belts.

9. Preparing for transport

△ — place guard over fingers and knife before raising cutterbar

△ — when lifting cutterbar, hold it under the bar, not by the fingers

△ — ensure transport rod is securely fastened so cutterbar will not drop down

- ensure all guards and check chains are correctly in place and mower is raised to correct transport height

△ — place the spare knife in a carrier and tie securely, so that it will not fall off during transport, and so that it will not become bent when lowering the mower.

The mower is now ready for work.

You should understand

- how to avoid risk of injury when handling knives
- how wear in the cutterbar leads to poor cutting and blocking
- the importance of a methodical approach to preparation
- why the manufacturer's instruction book needs to be carried with the machine always
- relevant agricultural safety legislation
- the importance of tetanus immunisation.

If you do not understand all these points, ask your instructor, manager or employer for advice.

Further experience

You will become proficient only by practice. Try to get experience as soon as you can by taking every opportunity offered by your employer or manager to follow up the training you have received.

Remember that the Board and your county agricultural college offer a wide range of courses in practical skills, together with associated knowledge which will help you understand more about the work you are doing.

31 OPERATING RECIPROCATING KNIFE MOWERS

Your aim

is to be able to:

— set up a rear-mounted reciprocating-knife mower to suit conditions in the field

— operate the mower in the most effective way

— deal with variations in crop and ground conditions

— correct any blockage or faults that occur.

You will need

— rear-mounted reciprocating-knife mower

— suitable tractor (min 26 kW)

— manufacturers' instruction books for mower and tractor

— normal hand tools in secure box

— oil can and grease gun containing clean lubricants of the correct type for the mower

— at least one spare knife in a carrying guard.

You should have

Received training or gained experience in:

— maintaining reciprocating-knife mowers — see trainee guide M.10.A.1

— storing machinery.

Note: △ = important safety point

1. Setting the mower for work

(a) *Before setting out for the field*

Ensure that mower is correctly hitched to the tractor and:

— frame is level from side to side

— mower is held on support chains (crossed) or position (height) control

— p.t.o. shaft halves are of correct lengths and shaft guards in place.

Carry out daily check on mower (all detailed maintenance should have been done):

— drive belts correct tension

— knife checked and sharp

— mower lubricated.

These items may need checking during the day's work. Find out from instruction book.

(b) *When at the field to be cut*

— swing cutterbar down into working position

△ — do not remove cutterbar guard until cutterbar is lowered

△ — always support the cutterbar by the bar itself and the outer shoe whilst lowering. Never put your fingers near the knife whilst raising or lowering the bar, as the knife moves during the lift

— set lift linkage as required by manufacturer to ensure proper 'float' of cutterbar during work. Refer to instruction book

— fit transport rod in working position or leave it safely by the hedge together with spare knife, cutterbar guard and toolbox.

Set the cutting height.

If you have not already been told the height, as a general rule cut to avoid the wet, brown bases of stems:

- there is little feed value so low down
- drying takes longer
- re-growth of the grass is slower.

Ground conditions also affect height:

- you can cut lower on stone-free ground than on uneven, stony ground.

Pitch (or angle) of cutterbar can be altered to assist in difficult conditions:

- normally cutterbar is level or pointing slightly down
- angle the cutter bar down to help lift up and run under badly flattened grass
- point the cutter bar up to run over stones where present.

Check swathboard and grass stick:

- swathboard should not be so loose that it is forced back by grass pressure but must be free to follow ground in work
- grass stick should be in forward holes for light, standing crops and further back for heavier, flattened grass.

Grass stick may have to be set after mowing has started:

- grass should run freely over stick during work. A second swathboard, if available, should be fitted and used when cutting the first round
- this makes cutting the backswath much easier, as there is less risk of blockages.

2. Operating the mower

The operations following depend on correct planning of the mowing.

There are basically two choices:

- cut all the field round and round

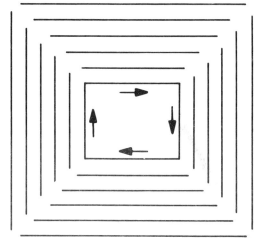

Field cut round and round

or:

- after cutting the headlands, work in 'lands' parallel with the longest straight side.

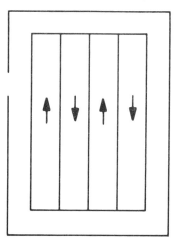

Field cut in lands

Cut the headlands first:

– drive clockwise round the field (with the hedge on your left)

– on the first round leave an uncut strip next to the hedge as wide as the cutterbar. This is called the backswath and is mown the other way round (anticlockwise)

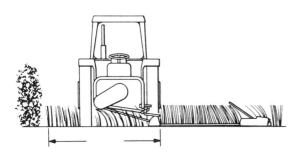

Backswath

– fit an inner swathboard, if available, when cutting the backswath

– engage p.t.o. drive when cutterbar is on the ground and smoothly increase speed to 540 rev./min.

– select a forward gear that will allow you to travel safely and to avoid obstructions, especially on the headlands. Travel slower during the first round to gain knowledge of the field.

Stop at corners when front of tractor is at the hedge or fence:

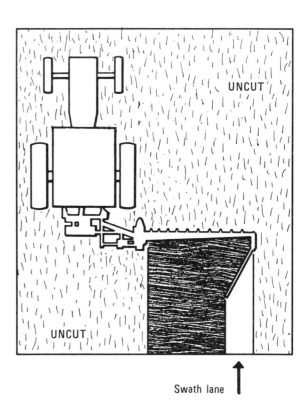

Swath lane

△ – allow more room between tractor and edge of field when cutting next to ditches. It is better to leave a strip of grass uncut than risk an accident.

Reverse round until lined up for next run.

On the following rounds, stop the tractor and raise the cutterbar when it is clear of grass in the swath lane:

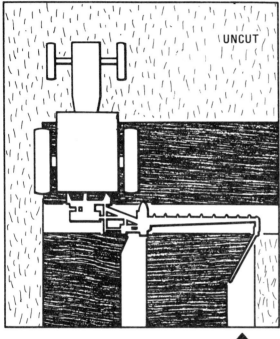

Swath lane

- reverse round and move forward, lowering cutterbar into the swath lane of the previous run then continue the next side

- raise cutterbar only enough to clear the swaths. Raising higher places a strain on p.t.o. shaft joints.

Cut backswath after two or three rounds. You will need some cut area in which to turn the tractor:

- do not leave backswath uncut for too long. It takes more drying than the rest due to shelter from hedge

- drive slowly while cutting the backswath as the inner shoe is very easily blocked by the grass already cut

- remember that hay and silage machinery cannot get right into the corners, so do not try to cut every blade of grass.

You must decide how wide to make the lands if you are cutting the field this way:

- usually work parallel with the longest straight edge of the field.

△ On sloping ground drive up and down the slope, not across it:

- when making the first cut through, try to keep parallel with the edge. Do not cut a triangle. Concentrate on keeping straight by aiming towards a point in the hedge.

Work up and down the new strip you have cut, travelling anti-clockwise to make it wider.

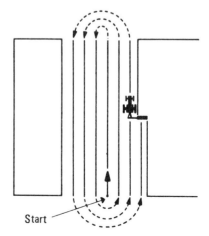

Start

When the land has become 20 – 30 m wide you will be spending too long travelling on the headlands:

– finish off the first block by travelling clockwise.

Small obstructions, e.g. trees can be driven round:

– cut out the curve in the row over the next two or three runs

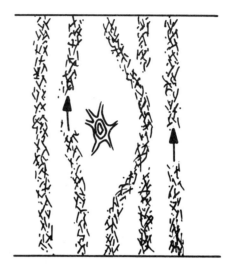

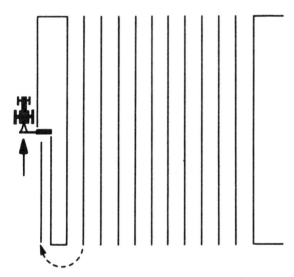

Large obstructions, e.g. ponds must have a headland cut around them to allow machines to turn.

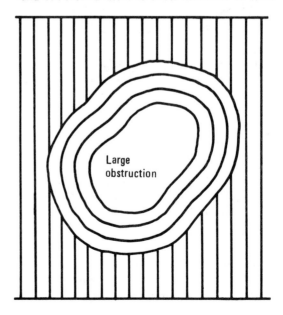

Now cut out a new strip and repeat the process until the field is finished.

Where there are flattened areas of grass, try to organise cuttings so that the grass points towards the cutterbar.

– grass falls over cutterbar and prevents blockage.

Ensure correct speed during cutting:

— always have engine revs set to give p.t.o. speed of **540** rev/min. Do not exceed this speed

— vary forward speed by use of tractor gears, not engine speed

— travel at a safe and comfortable speed

△ — you must always have full control over steering

Always steer so that inner shoe runs in the swathlane close to uncut grass:

— this ensures that cutterbar always cuts its full width

— when right-hand curves get too tight to go round without stopping, make them into angled corners like the corners of the field

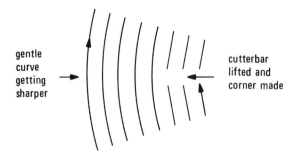

right-hand curves

left hand curves get wider so you can follow the curve each time round.

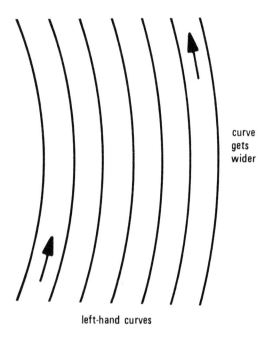

left-hand curves

Always be alert for blockages when mowing:

— blockages are far more likely with a badly maintained cutterbar

— to clear a blockage, reverse with cutterbar on the ground, still running, then drive forward fairly quickly. The lump of grass should pass over the cutterbar

— take care that swathboard does not dig into ground when reversing

— blockages may also be cleared by raising the cutterbar and dropping the grass onto the uncut crop. You can now mow underneath.

If these methods fail you will have to clear the blockage by hand:

△ — disengage p.t.o., stop tractor engine and apply handbrake

△ — never ever touch any part of the mower while it is running

— pull lumps of grass away and spread them out on the cut swaths

△ — take care when pulling grass from the sections or fingers

— use this chance to inspect cutterbar

— replace knife or re-adjust if sections are damaged or there is too much free play in the knife.

Be prepared to change the knife at least halfway through the day's cutting even if no damage has been done:

— in tough, heavy crops the knife may need changing even more frequently

— it evens out the wear between two knives

— the frequent blockages are a sign that the spare knife should be used (there may be other reasons also).

3. Preparing for next day's work

— refit knife guard for transport

— lubricate the mower as per manufacturers' instructions

— re-sharpen both knives

— clean fingers and cutterbar

— fit one knife and check free play

— re-adjust wear plates and clips if too slack

— fold cutterbar up into the transport position and lower the mower onto the ground overnight.

4. Laying-up the mower out of season

Carry out the procedure as in trainee guides on laying-up machinery out of season.

Make sure that knives are well oiled and are stored in a safe place.

Place suitable protection over the cutting edges.

Make sure the mower is safe when left. If necessary place blocks under parts of the frame to prevent it falling.

You should understand

- the importance of planning your mowing

- why correct speed is important

- the causes of blockages of cutterbar

- the need to change knives regularly

- relevant safety legislation

- the importance of being fully immunised against tetanus.

If you do not understand all these points, ask your instructor, manager or employer for advice.

Further experience

You will become proficient only by practice. Try to get experience as soon as you can by taking every opportunity offered by your employer or manager to follow up the training you have received.

Remember that the Board and your County Agricultural College offer a wide range of courses in practical skills, together with associated knowledge, which will help you understand more about the work you are doing.

Fault Chart

Refer to this fault chart to help you out while mowing:

Fault	Cause	Remedy
Break-back operates	Obstruction	Reverse to re-engage. Remove obstruction or go round
" + blockage	Spring not tight enough	Tighten spring until break-back just holds
	Knife speed too slow, or forward speed too fast	Ensure correct p.t.o. speed or change gear down
	Not enough cutterbar lead	Re-set lead
" + blockage	Blunt sections, or broken sections or knife lifting off ledger plates	Fit spare knife. Re-adjust clips and wear plates
" + blockage	Knife register wrong—crop dragging	Re-set register
" + skid marks	Cutterbar lift wrongly set—too much weight on skids	Re-set lift mechanism so cutterbar floats in work
" + soil over cutterbar	Fingers digging in	Raise fingers until just below horizontal
Break-back fails to operate when large blockage or other fault occurs	Spring set too tight	Release spring until break-back just holds in work. Check cutterbar for damage
P.t.o. drive clutch slips (knife slows) or ratchet clutch works (clacking noise)	Blockage	Remove blockage. Trace cause (see above)
	Clips, wear plates too tight or damaged knife	Re-adjust clips, wear plates or fit spare knife
Broken driveshaft	Severe damage to knife or cutterbar	Repair damage, fit spare knife and re-adjust.